MICROBIOLOGY

About the Author

John J. Lee is Professor of Biology at City College of City University of New York, where he teaches general microbiology, marine microbiology, microbial ecology, and radiobiology. He received the degrees of B.S. from Queens College of City University of New York, M.A. from the University of Massachusetts at Amherst, and Ph.D. from New York University. Postgraduate studies took place at Oak Ridge Institute of Nuclear Studies and the University of California at Berkeley. Dr. Lee has conducted research at Haskins Laboratories, the American Museum of Natural History, and City College in New York City; the Marine Biological Laboratory, Woods Hole, Massachusetts; the Heintz Steinitz Marine Biology Laboratory, Elat, Israel; and Key Largo Sound, Florida.

Dr. Lee is widely known for his research on marine protozoa, microbial food webs, and marine benthic unicellular algae, and for his gnotobiotic studies of micrometazoa (meiofauna). He has written chapters for many books and has contributed numerous papers to scientific journals; his popular articles have appeared in *Natural History* magazine. In addition to holding various offices in the Society of Protozoologists, he has served as Secretary General of the V International Congress on Protozoology and Secretary of the International Commission on Protozoology of the IUBS (International Union of Biological Sciences).

MICROBIOLOGY. Copyright © 1983 by John J. Lee. All rights reserved. Printed in the United States of America. No part of this book may be used or reproduced in any manner whatsoever without written permission except in the case of brief quotations embodied in critical articles and reviews. For information address Harper & Row, Publishers, Inc., 10 East 53rd Street, New York, N.Y. 10022. Published simultaneously in Canada by Fitzhenry & Whiteside Limited, Toronto.

FIRST EDITION

Designer: C. Linda Dingler

Library of Congress Cataloging in Publication Data

   Lee, John J.
   Microbiology

   Bibliography: p.
   Includes index.
   1. Microbiology.   I. Title.
QR41.2.L43   1982   576   80–8418
ISBN 0–06–460183-8 (pbk.)   AACR2

---

83 84 85 86 87 10 9 8 7 6 5 4 3 2 1

MICROBIOLOGY

JOHN J. LEE

BARNES & NOBLE BOOKS
A DIVISION OF HARPER & ROW, PUBLISHERS
New York, Cambridge, Philadelphia, San Francisco
London, Mexico City, São Paulo, Sydney

CONTENTS

	Preface	vii
1.	Introduction to Microbiology	1
2.	Tools and Methods of Microbiologists	15
3.	Procaryotic Cell Structure	37
4.	Microbial Growth	49
5.	Microbial Nutrition and Metabolism	59
6.	Viruses	91
7.	Genetics of Microorganisms	103
8.	Symbiosis	114
9.	Microbial Activities in the Biosphere	120
10.	Applied and Industrial Microbiology	129
11.	Microbial Pathogens and Diseases of Humans	140
12.	Summary of Procaryotic Groups	169
13.	Eucaryotic Microorganisms	206
	Supplementary Reading	253
	Index	265

PREFACE

Microbiology is a multifaceted discipline which is taught in American colleges and universities with a wide range of emphasis on individual topics. Some courses concentrate on the biology and evolution of microorganisms, others are medically oriented, still others are food or industry related. No one text covers all topics with the same vigor. This book is an overview of microbiology, presented in a concise and easily understood manner. As such it can be used in conjunction with a variety of courses and texts to facilitate study and review. The student should be aware that microbiology is a rapidly advancing science and that many classroom lectures will present more up-to-date information than will be found in this and other texts now in print.

This book should also serve as a brief introduction to the fascinating world of microbes. Many of the articles cited at the end of the book were selected on the basis of their wide appeal as well as for the additional information they can provide.

I am extremely grateful to my wife, Judith, for typing the manuscript and proofreading the many revisions. Special thanks are due to Mrs. Gertrude Fisher of the City College of CUNY, who made most of the line drawings. I am also indebted to the many undergraduate students in microbiology at the City College and Bucknell University who provided constructive criticism of early drafts of the manuscript and thus helped me achieve a greater clarity of presentation.

Chapter 1

INTRODUCTION TO MICROBIOLOGY

It is difficult to imagine life here on earth without microorganisms. Microbes are found in every conceivable environment: at the bottom of the world's oceans (low temperature and very high pressure), in thermal springs (highest temperature for life forms), in salt lakes and dead seas (highest osmotic pressures tolerated by life forms), and in tropical sands, deserts, and soils (extremes of radiation, temperature, and desiccation). These tiny creatures affect every facet of life. They play essential roles in the production of foodstuffs, vitamins, and chemicals, and in the decay and recycling of every biologically important mineral. In the century since Pasteur, microbiology has grown from a flourishing applied science to one also concerned with basic cellular and molecular processes.

THE NATURE AND CLASSIFICATION OF MICROORGANISMS

Microorganisms, or microbes, are organisms that are too small to be seen easily by the unaided eye. They are roughly 1 mm (1000 μm) or smaller in size. The smallest microbes, certain types of viruses, are only 20 nm (2×10^{-5} mm).* Microorganisms belong to diverse taxonomic groups which include bacteria, protozoa, algae, fungi, and viruses. Although many groups of small metazoan animals such as rotifers, nematodes, and microcrustacea fall into the

---

*The unit of length for fine structure is the nanometer (nm). One nm equals 0.001 or 10^{-3} micrometer (μm).

upper part of the microbial size range, recent common practice has been to place them in a separate category, the *meiofauna,* defined as organisms ranging from 0.05 to 0.5 mm in size. The larger protozoa—for example, some ciliates, foraminifera, and radiolaria—also fit into this category.

Major Categories of Microorganisms. On the basis of their type of organization, microorganisms can be conveniently divided into three major categories: viruses, procaryotes, and eucaryotes (see Table 1.1).

VIRUSES. Viruses are extremely small ($\leq 0.300 \mu m$) subcellular obligate intracellular parasites consisting of a core of nucleic acid (either RNA or DNA but not both) usually enclosed in one or two protein coats. Viruses do not grow or reproduce but are replicated by host cells whose energy, metabolites, and synthetic abilities are diverted for the process. Viruses are not cultivable on nonliving laboratory media nor are their activities observed in the absence of either procaryotic or eucaryotic cells. Viruses have no—or very few—enzymes. They are unable to generate or release energy.

PROCARYOTES. Bacteria and cyanobacteria (blue-green algae) belong to this group. The distinctive characteristic of procaryotic cells is the absence, throughout the cell cycle, of a membrane that separates the nuclear material from other cell components. Most of the DNA of procaryotes is organized on a long continuous circular thread, a type of chromosome, properly termed a *genophore.* A genophore is less compact and lacks the basic protein (histones) associated with the chromosomes of eucaryotes. Procaryotic cells also lack mitochondria, chloroplasts, endoplasmic reticula, Golgi, and centrioles; these organelles are found only in eucaryotic cells. The flagella of those bacteria that have them are constructed of spheroid subunits in a tightly wound helix quite different from the long 9 + 2 fibrillar structures typical of eucaryotes. The ribosomes of procaryotes are smaller (70S) than those found in eucaryotes (80S).* Not all procaryotic cells have cell walls, but in most of those that do, the wall has a unique kind of strengthening organic polymer known as *murein* (peptidoglycan). Another chemical peculiarity of nearly all procaryotic cells is a negative one—they lack sterols. Sterols are a class of lipids, all of which have a distinctive

---

*S stands for the Svedberg unit, a relative measure of the rate of sedimentation of cellular components or molecules during centrifugation.

Table 1.1. Major Differences between Viruses, Eucaryotes, and Procaryotes

	VIRUSES	PROCARYOTES	EUCARYOTES
Nuclear membrane	None	None	Present
DNA organization	Genophore. Single loop of double- or single-stranded DNA or RNA. Not complexed with histones.	Genophore/nucleoid. Single continuous loop of double-stranded DNA. Not complexed with histones.	In one or many chromosomes. Complexed with histones. Some organisms may have more than one set of chromosomes.
Division of DNA	No mitosis. Replication of nucleic acids by host.	No mitosis. Genophore associated with cell membrane during replication.	Mitosis. In most groups, nuclear membrane breaks down and microtubular spindle is formed.
Genetic recombination	Usually involves small fragment of entire set of genes.	Usually involves small fragment of entire set of genes (heterozygotes formed). No meiosis. Many parasexual means of recombination.	Sexual reproduction in most higher forms. Whole sets of chromosomes involved. Meiosis unknown in many microbial forms.
Plasma membrane	None	Present; lacks sterols (except in some groups). Respiratory enzymes associated with plasma membrane or membranous infoldings from it.	Present; sterols usually present. Respiratory enzymes associated with mitochondria.
Membrane-enclosed organelles	None	Cyanobacterial thylakoids	Mitochondria Chloroplasts Endoplasmic reticulum Golgi Lysosomes Peroxisomes Glyoxysomes Flagella and cilia (9 + 2 organization)

condensed ring structure. They are important components in eucaryotic cell membranes.

EUCARYOTES. These cells are characterized by a nuclear membrane that separates the nucleus from the cytoplasm at most stages of the cell cycle. The chromosomes of eucaryotes are generally more numerous and compact than those of procaryotes and associated with basic proteins (histones). Eucaryotes have more DNA (>100×) than procaryotes. Eucaryotic cells typically have an endoplasmic reticulum, mitochondria, Golgi, and lysosomes, and may have centrioles, flagella of the 9 + 2 design, and chloroplasts. Protein synthesis takes place on 80S ribosomes. Eucaryotic cells synthesize or require sterols for cell membranes. The cell walls of various eucaryotes are strengthened by cellulose, silica, calcite, chitin, and other structural material but not by murein. Nuclear division in eucaryotes is a process called mitosis. Many eucaryotic cells have the ability to undergo pinocytosis, phagocytosis, amoeboid movement, and cytoplasmic streaming, phenomena unknown in procaryotes.

Classification of Microorganisms. The classification of living organisms, or *taxonomy,* is a conceptual science which looks at diversity in organisms and attempts to group them according to some logical plan. The organisms within groups or categories called *taxa* share common attributes and are believed to be evolutionarily related to one another.

The highest taxonomic unit is the kingdom, followed by phylum, class, order, family, genus, and species. The latter two taxonomic units, genus and species, are used to refer to particular organisms (e.g., *Homo sapiens, Escherichia coli*). This practice is known as the binomial system of nomenclature. (The Latin names are underlined in manuscript form and italicized in print. The genus is capitalized; the species name is lower case.)

Microorganisms are particularly difficult to classify. There are several reasons for this: (1) they are a very diverse group, their only common characteristic being their small size; (2) breeding is primarily asexual; (3) the fossil record for most forms is scanty or nonexistent, making evolutionary changes difficult to trace; and (4) as yet too few species have been examined by newer taxonomic tools such as the transmission and scanning electron microscopes (TEM and SEM).

THE PROBLEM OF KINGDOMS. Over a century ago, biologists recognized that microorganisms were not appropriately placed in either the animal or plant kingdom. In 1866 the German biologist Haeckel created a third kingdom, the *Protista,* for unicellular organisms. As our knowledge of cellular structure and evolution increased, particularly in the past 40 years, it became obvious that a more comprehensive taxonomic scheme was needed to separate viruses, procaryotes, and the diverse types of eucaryotic microorganisms from the animal and plant kingdoms.

Today one of the most widely accepted schemes (although it still does not have a place for viruses) is Whittaker's five-kingdom system (Fig. 1.1.) based on three levels of organization: procaryotic, eucaryotic unicellular, and eucaryotic multicellular. The five kingdoms are as follows: (1) *Monera:* procaryote; (2) *Protista:* eucaryote unicellular; (3) *Plantae:* eucaryote multicellular photosynthetic; (4) *Animaliae:* eucaryote multicellular with ingestive nutrition; and (5) *Fungi:* eucaryote multicellular with absorptive nutrition. The three kingdoms in Whittaker's scheme which contain microorganisms are Monera, Protista, and Fungi. They are set forth in Table 1.2.

Table 1.2. Whittaker's Classification of Microorganisms

1. *Kingdom Monera* (procaryote cells)
Organisms lacking nuclear membranes, mitochondria, chloroplasts, endoplasmic reticula, Golgi, centrioles. Unicellular or primitively colonial. Predominant nutritional types are absorptive, some groups are photosynthetic. Reproduction primarily asexual by fission or budding. Genetic recombination through various modes of parasexual phenomena (see chapter 7). Nonmotile or motile by means of simple flagella and gliding.

Myxomoneran Branch: With gliding motility and related nonmotile forms
    Phylum Cyanophyta (cyanobacteriae)   blue-green algae
    Phylum Myxobacteriae   gliding bacteria

Mastigomoneran Branch: With simple flagella and related nonmotile forms
    Phylum Eubacteriae   true bacteria
    Phylum Actinomycota   mycelial-forming bacteria
    Phylum Spirochaetae   spirochetes

2. *Kingdom Protista* (basically unicellular, eucaryotic cells)
Organisms having nuclear membranes, mitochondria, endoplasmic reticula with 80S ribosomes, Golgi, lysosomes. Primarily unicellular, but some groups are advanced colonial forms with cellular differentiation. True sexual reproduction known in most groups but reproduction is primarily asexual by binary fission, budding, and multiple fission. Diverse modes of nutri-

Table 1.2. Continued

tion including photosynthetic, absorptive, and ingestive species and some which have combination of all 3 (or 2 of the 3) modes. If motile, motility by true flagella (9 + 2 pattern), gliding, amoeboid movement, and other means.

Green Branch: Predominantly green and pigmented

Phylum Euglenophyta	*Euglena* and related organisms
Phylum Eustigmatophyta	eustigmatophyte flagellates with scaly flagella
Phylum Chlorophyta	*Chlamydomonas, Volvox,* and related organisms

Heterokont Branch: Golden pigmented forms with one tinsel or hairy flagellum

Phylum Bacillariophyta	diatoms
Phylum Chrysophyta	golden algae
Phylum Xanthophyta	xanthophyte flagellates

Unrelated Algal Phyla on Separate Branches

Dinophyta (Pyrrophyta)	dinoflagellates
Cryptophyta	cryptomonads
Haptophyta	haptophytes

Protozoan and Lower Fungal Branch: Nonchloroplast bearing, usually motile eucaryotic cells

Plylum Zoomastigina	unpigmented flagellates
Plylum Sarcodina	amoebae and other sarcodines
Phylum Ciliophora	ciliates and suctorians
Phylum Sporozoa	sporozoans
Phylum Cnidosporidia	cnidosporidians
Phylum Chytridiomycota	chytrids
Phylum Myxomycota	myxomycetes, acellular slime molds
Phylum Acrasiomycota	cellular slime molds
Phylum Plasmodiophoromycota	plasmodiophores

3. *Kingdom Fungi.* Primarily multinucleate, nonphotosynthetic eucaryotic organisms with cell walls but limited somatic tissue differentiation and often with characteristic sexual and asexual life cycles not similar to those found in the plant kingdom. Primarily nonmotile, living in or on food supply.

Phylum Zygomycota	aseptate molds
Phylum Ascomycota	yeasts and septate molds
Phylum Basidiomycota	rusts, smuts, toadstools, and mushrooms

Recently it has been argued (Leedale and others) that lumping all of the eucaryotic unicellular organisms into one kingdom (Protista) does not make natural sense, since many groups seem to be evolutionarily unrelated, and since in some groups there are natural and gradual progressions from unicellularity to multicellular differentiation. Schemes that reflect this concern are shown in Figs. 1.2 and 1.3.

Many microbiologists feel that the scheme presented in Fig. 1.3, with its multiple kingdoms, is *too* extreme. An international commission of protozoologists recently (1980) decided that the protozoa were a kingdom comprised of seven phyla. Their new classification scheme includes all or parts of 10 groups of organisms that were considered separate kingdoms in the scheme in Fig. 1.3. In compromise schemes (e.g., Fig. 1.2), some groups seem to fit into more than one major category. For example, dinoflagellates can be considered either protozoa, algae, or a separate kingdom, depending on the taxonomic approach used.

THE PROBLEM OF SPECIES. The species concept as applied by biologists to higher organisms is difficult to apply at the microbial level. A vertebrate biologist would define a species as a group of organisms which actually, or potentially, breed with each other and not with organisms outside the group, to produce fertile offspring. Since asexual reproduction is the most common reproductive mode in microorganisms and the breeding systems are poorly known or do not exist, other criteria are used to define microbial species. Morphological criteria are used to separate most eucaryotic microbial species. So many different bacteria share the same morphological characteristics, however, that physiological, nutritional, biochemical, metabolic, antigenic, genetic, and pathogenic criteria are also used. Classical (traditional, Linnaean) taxonomic approaches have been built on hierarchical systems in which organisms are separated by key (very important) characteristics. For example, we easily separate animals with and without backbones into vertebrates and invertebrates, and warm blood animals with four-chambered hearts into birds (Aves) if they have feathers and mammals if they have hair.

For the most part bacteria are classified this way. The taxonomic approach used by most microbiologists is classical, as illustrated by *Bergey's Manual of Determinative Bacteriology*, the standard reference work in the field (see Chapter 12). In recent years microbiologists have begun to question traditional taxonomic approaches.

8   *Introduction to Microbiology*

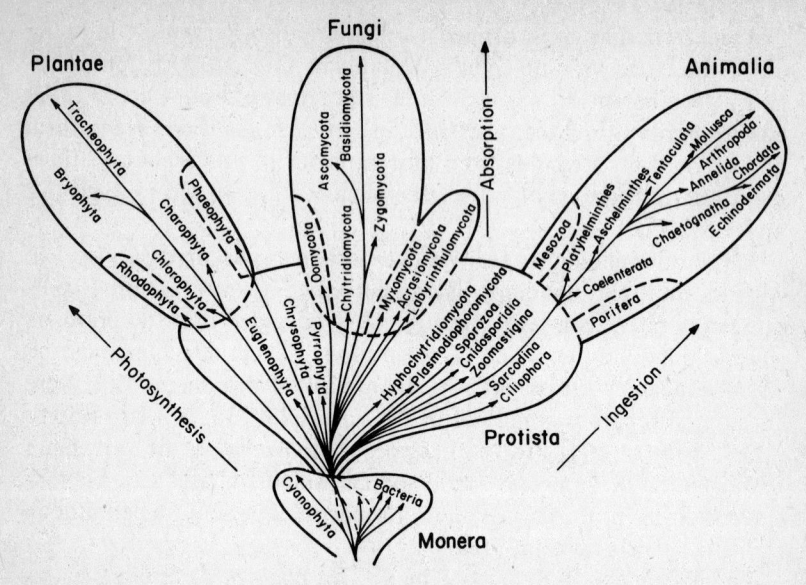

Fig. 1.1. Whittaker's five-kingdom system of classification.

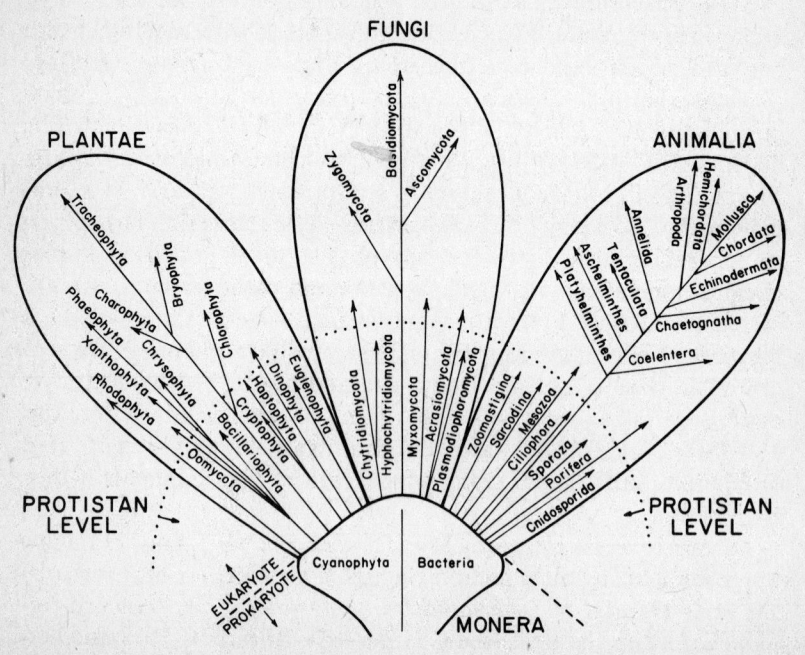

Fig. 1.2. Compromise classification scheme.

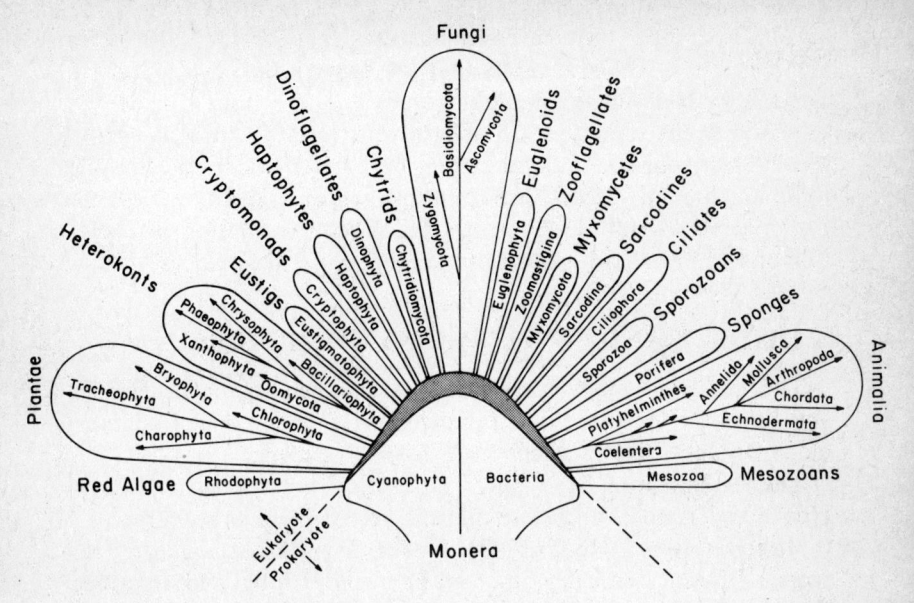

Fig. 1.3. Leedale's multikingdom system of classification.

Complications at the microbial level arise because some "important" characters are highly variable, and it is difficult to decide what are the most important characters to use. For these reasons, *numerical taxonomy* or *Adansonian taxonomy* has become increasingly popular among microbiologists. To use numerical taxonomy, every characteristic and bit of information about an organism is treated equally. No character is more important than any other. A computer program is used to compare the characteristics of each strain of bacteria with all others being considered. The computer program groups strains according to the number of characteristics they share and separates strains that have few common characteristics. Latin binomial names (genus and species) are then given to each group with common properties.

The Early Evolution of Microorganisms. Although the earth was probably formed about 4.5×10^9 years ago, the oldest rocks that have been found so far are only 3.5×10^9 years old. Fossil communities of bacteria and blue-green "algae" (cyanobacteria) have been found in sedimentary rocks nearly as old. The fossil record from this early period of life is extremely sparse. However, pre-Cambrian sediments indicate that by 8.5×10^8 years ago, eucaryotic microorganisms, archetypal algae, were well established and relatively diverse. These forms may have originated as early as 1.4×10^9 years ago. Current evidence suggests that the bacteria and cyanobacteria arose in an atmosphere devoid of oxygen and at a time when strong solar ultraviolet radiation was able to penetrate the atmosphere. As these forms evolved, the atmosphere changed from a reducing one to an aerobic one, because sufficient oxygen was released by cyanobacterial photosynthesis. This, in turn, paved the way for the evolution of eucaryotes, which, with rare exceptions, are aerobes.

HISTORY AND DEVELOPMENT OF MICROBIOLOGY

The names of many investigators are linked to the discoveries and controversies of the past three centuries during which microbiology developed into a science.

The Discovery of Microbes. The earliest microscopists, Robert Hooke and Anton Van Leeuwenhoek, were contemporaries.

ROBERT HOOKE (1635–1703). Hooke, using a compound microscope with reflected light, observed filamentous fungi and described them in his famous *Micrographia*.

ANTON VAN LEEUWENHOEK (1632–1723). With his powerful lenses, Van Leeuwenhoek looked everywhere with an inquisitive eye. He described swarms of tiny creatures he called animalcules from his saliva and mouth scrapings, ditch water, pepper infusions, urine, and other sources.

The Spontaneous Generation Controversy. From our twentieth-century perspective it is hard to appreciate the culture and background of persons living two to three centuries ago. In the absence of our knowledge of evolution and genetics, scientists of the seventeenth century argued whether life could arise spontaneously from decaying organic matter, fertile mud, and other media.

FRANCESCO REDI (1626–1679). Redi, who was doubtful about spontaneous generation, showed that maggots did not develop in putrefying meat if the meat was placed in a gauze-covered container. Flies attracted to the meat laid eggs on the gauze, but no maggots appeared in the meat as it decayed.

LOUIS JABLOT (1645–1723). Jablot was one of the first experimentalists to tackle the question in regard to the recently observed animalcules. To test the source of microorganisms in a hay infusion, he boiled some hay and placed half in a baked and closed vessel and the other half in an open vessel. In a few days the open vessel was swarming with microorganisms; none appeared in the closed vessel as long as it was sealed. Jablot concluded that the hay infusion once freed of life by heat was incapable of giving rise to new life spontaneously.

JOHN NEEDHAM (1713–1781). Needham got entirely different results with boiled mutton broth. Microorganisms appeared in vessels sealed with corks as well as in those open to the air.

LAZZARO SPALLANZANI (1729–1799). Spallanzani repeated Needham's work but heated the broth longer, and instead of using a cork, sealed the flasks hermetically with a flame. No microorganisms arose in the sealed flasks. Needham claimed that Spallanzani had destroyed the "vital force" by overheating the flasks. Spallanzani answered these objections by breaking the flasks open and showing that animalcules would then develop. Needham's counterargument was that it was merely the lack of air that had kept microorganisms from developing in the broth. More than half a century passed until the issue was resolved.

THEODOR SCHWANN (1810–1882) and FRANZ SCHULZE (1815–1873). Schwann and Schulze refuted Needham's counterargument by passing air freely to previously heated meat or hay infusions but only after it was passed through a red-hot glass tube or a strong acid or base solution. Experimental flasks remained clear and control flasks turned cloudy. Critics of these experiments suggested that the life-giving power of the air had been destroyed. In 1850 *Van Dusch* and *Schroeder* changed the experimental apparatus by using cotton to filter out the microorganisms.

LOUIS PASTEUR (1822–1895). Pasteur ended the controversy by using a very long gooseneck extension to his flasks. The air freely entered Pasteur's flasks but the microorganisms settled out in the bend of the long neck before entering the sterile broth.

JOHN TYNDALL (1820–1893). Tyndall explained why some experimental infusions produced microorganisms even after prolonged boiling. In 1876, using a special dust-free apparatus, he concluded that an inactive form of bacteria (later known as a spore) is carried in the air on motes of dust. The significance of these dormant forms was soon recognized independently by Robert Koch and Ferdinand Cohn.

Fermentation. Louis Pasteur also played a key role in the acceptance of the microbial theory of fermentation. The idea that biological activity was responsible for fermentation had been proposed, but found no support from the leading chemists of the day. Pasteur showed that the fermentation of beers and alcohols was not a simple chemical process but was in fact brought about by the action of living yeasts. He also demonstrated that various types of fermentive end products were produced by specific kinds of microbes. During his studies of the making of wine and beer, Pasteur found that the "diseases" of wine which brought so much economic loss to the industry were caused by undesirable fermentation by contaminating microbes. He also discovered that the undesirable microbes could be eliminated without damaging the quality of the wine by heating it to 60 to 65°C. This process of heating, known as *pasteurization*, is widely used in the food-processing industry, particularly for destroying disease microbes that might be present in milk.

The Germ Theory of Disease. During the first half of the nineteenth century, many investigators reported the co-occurrence of microorganisms of one kind or another and various diseases. In 1840 Jacob *Henle* expressed the germ theory almost exactly as we understand it today. However, without experimental proof, Henle's ideas were not readily accepted.

KOCH'S POSTULATES. In 1882, Robert *Koch* (1843–1910) made known his now famous postulates. Before any organism can be accepted as the cause for a particular disease all the following steps must be carried out:

1. The particular microorganism must be found in every case of the disease and not in healthy animals.
2. It must be isolated from the diseased animals and grown in the laboratory apart from all other organisms.
3. It must reproduce the symptoms of the disease when inoculated by itself into a healthy susceptible animal.
4. The same organism must be found again in the diseased animals and isolated again in laboratory culture.

PURE CULTURE METHODS. Koch's postulates assume that pure cultures are obtainable. Koch isolated many pathogenic microorganisms including the tubercle, cholera, and anthrax bacilli, and he and his collaborators were responsible for developing many of the isolation and culture techniques we use today. In his laboratory, gelatin and then agar were used as solidifying agents for microbial media. Slant cultures, and the streak and pour plate methods for isolating and counting bacterial colonies originated there also. R. J.

Petri, a student of Koch, developed the shallow flat-bottomed dishes with covers that are so widely used today.

Immunity, Chemotherapy, and Antibiotics. The idea that once an individual contracted a specific disease he would not contract it again existed in many cultures. The ancient Chinese deliberately inoculated healthy people with sores from people with mild cases of smallpox. In 1796 Edward *Jenner* introduced and popularized in England the practice of vaccination for smallpox with a cowpox vaccine. Pasteur developed vaccines against fowl cholera, rabies, and anthrax. As his work progressed, he discovered four basic methods of attenuating (weakening) organisms for use in preparing vaccines. In 1884, Elie *Metchnikoff,* a student of Koch, proposed the cellular theory of immunity. Working with the water flea, he observed amoeboid cells ingesting disease-causing yeasts, a process he called phagocytosis.

Paul *Ehrlich,* another of Koch's students, is best known for his searches for "magic bullets"—chemicals that would attack microbial cells but not animal cells. His discovery of an organic arsenical that was effective in destroying the syphilis spirochete in the human body laid the groundwork for the modern era of chemotherapy. Ehrlich also advanced the theory of humeral immunity, which suggested that soluble substances in body fluids were responsible for immunity. More recently (1960) MacFarlane *Burnet* and Peter *Medawar* won Nobel Prizes for setting forth the immunological factors involved in the acceptance or rejection of grafts.

Chemotherapy was advanced by the discovery of the therapeutic value of a compound called prontosil by Gerhard *Domagk* in 1931 (Nobel Prize 1939). From the basic formula sulfanilamide came many derivatives, known as the sulfa drugs, that were found to be effective against bacterial infections.

The modern antibiotic era began in 1928 with Alexander *Fleming's* discovery of penicillin. He shared the 1945 Nobel Prize in Physiology or Medicine with Ernst *Chain* and Howard *Florey,* who demonstrated the medical use of penicillin. After Selman *Waksman's* discovery of streptomycin (Nobel Prize in 1952), the discovery of many other antibiotics followed quickly.

Discovery of Viruses. Although Pasteur worked with rabies viruses, he had no idea the agents he was working with were so small. The first disease clearly demonstrated to be caused by a submicroscopic agent was tobacco mosaic disease. D. *Iwanowsky* (1892) passed the causative agent through porcelain filters small enough to retain all bacteria. The filtrate was able to cause the disease. M. *Beijerinck* (1899) confirmed Iwanowsky's work and proposed the existence of a new class of organisms. Friedrich *Loeffler* and P. *Forsch* (1898) discovered that the causative agent of foot-and-mouth disease, the first animal disease attributed to a virus, was too small to be seen with an ordinary microscope and would not grow on ordinary laboratory media. The first bacterial viruses were discovered independently by F. W. *Twort* (1915) and F. *d'Herelle* (1917). In 1935 Wendell *Stanley* first isolated, purified, and crystalized the tobacco mosaic virus. *Hershey* and *Chase* (1952) were the first to demonstrate that viral nucleic acid carries genetic information and that during infection only the viral nucleic acid penetrates the cell and initiates the replicative cycle. *Enders, Robbins,* and *Weller* (Nobel Prize 1954) found that polio virus could be cultured in the kidney cells of monkeys, thus laying the groundwork for the later development of polio vaccines. Peyton *Rous* won a Nobel Prize (1966) for his demonstration that a virus could cause cancer in an experimental system.

Microbial Genetics. Many notable individuals have been involved in this recent and rapidly unfolding field. One of the milestones was the 1941 discovery and use of nutritional mutants of the mold *Neurospora* by George *Beadle* and Edward *Tatum* (Nobel Prizes 1958). They correlated enzyme differences with genetic differences and concluded that genes controlled enzyme activity. Their theory, one gene-one enzyme, laid the groundwork for much of our present genetic research. O. T. *Avery,* C. M. *MacLeod,* and M. *McCarty* were not the first to study transformation in bacteria, but working with pneumococci in 1944, they demonstrated that DNA was the substance responsible. In 1958, Joshua *Lederberg, Beadle,* and *Tatum* shared a Nobel Prize for their clever demonstration of recombination in bacteria and the discovery of transduction. Both discoveries paved the way for the mapping of fine structure and functioning of the bacterial genophore. At the Pasteur Institute, a team of Francois *Jacob,* Andre *Lwoff,* and Jacques *Monod* uncovered the details of the complicated process by which genes regulate the synthesis of enzymes and other proteins (Nobel Prize 1965).

Chapter 2

TOOLS AND METHODS OF MICROBIOLOGISTS

A variety of techniques and optical tools are employed to separate, isolate, identify, and observe microorganisms and their activities. Each technique and instrument offers some specific advantage for the demonstration of microbial characteristics, and their mastery is part of the training of every microbiologist.

MICROSCOPY

Most microorganisms are so small that they are invisible to the unaided eye. A magnified image can be produced by two different means: by light waves, as in the light (optical) microscope, and by a beam of electrons, as in the electron microscope.

Optical Microscopy. Microbiologists commonly use six different types of light microscopy: (1) bright-field, (2) dark-field, (3) ultraviolet, (4) fluorescence, (5) phase-contrast, and (6) interference.

BRIGHT-FIELD. The most widely used optical instrument is the bright-field microscope (Fig. 2.1). Microscopes of this kind require the least amount of training to use and generally produce a useful magnification up to 1000×. If we try to increase magnification beyond this point, the image gets fuzzy. The basic limitation is not entirely a matter of magnification and lens quality, but a phenomenon related to the wavelength of light used, called *resolution* or *resolving power*. Resolution is the ability to distinguish two adjacent points as separate and distinct from one another (Fig. 2.2). In theory, a light microscope can resolve objects as small as 0.2 mm, but in actual practice, resolution varies from 0.3 to 5.0 μm. Resolv-

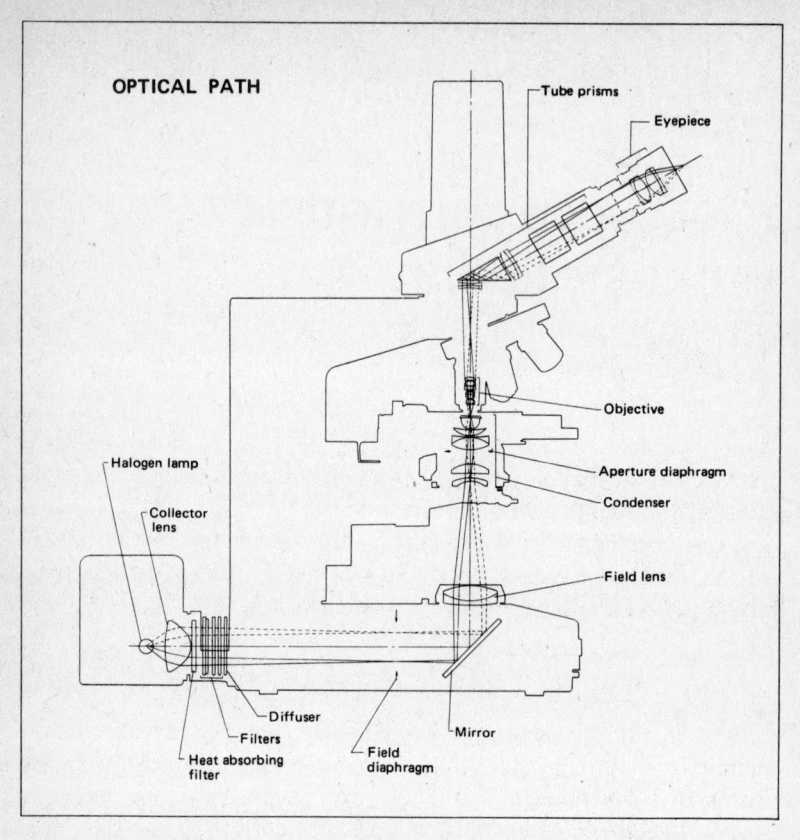

Fig. 2.1. Bright-field microscope. Courtesy of Nikon.

ing power depends on the wavelength of light used and on a characteristic of lens systems known as numerical aperature (NA).

$$\text{resolving power} = \frac{\text{wavelength of light}}{\text{NA}_{\text{objective}} + \text{NA}_{\text{condenser}}}$$

Numerical Aperture (NA). Numerical aperture is an expression of the angle of light gathered by a lens, and is affected by the refractive index of the medium—air or oil—between the front lens and the coverslip. To attain the high resolution necessary to see very small objects such as bacteria, oil-immersion lenses are used. Oil has a higher refractive index than air; thus it permits lights rays to pass almost unbent between the slide and the front lens of the objective.

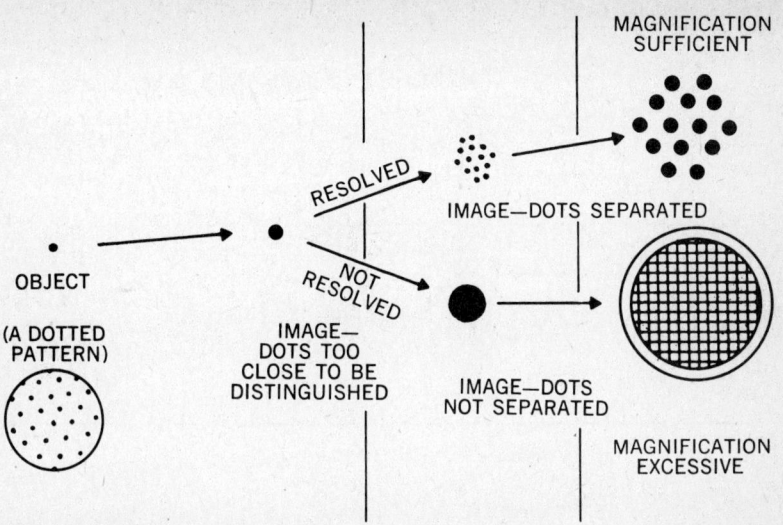

Fig. 2.2. Resolution.

Spherical and Chromatic Aberration. Two additional lens qualities—chromatic and spherical aberration—affect the quality of the image.

Chromatic aberration is a blurring of the image due to the fact that light of shorter wavelengths (blue) focuses closer to the lens system than light of longer wavelengths (red) (Fig. 2.3). In high-quality microscopes, this condition is overcome by combining lenses manufactured from glasses with different optical qualities. Depending on the combinations, such color-corrected compound lens systems are known as *achromatic, apochromatic, fluorite,* and *neofluor.*

Spherical aberration is blurring of the image due to differences in the length of the path of light in different parts of a simple lens. In a biconvex lens, the light passing through the edge of the lens will focus closer to the lens than light passing through the center. When not corrected, the object will be in sharp focus in the center of the optical field and blurred at the edges. Spherical aberration is corrected by combining appropriately ground lenses into compound systems. Particularly well-corrected lenses are known as *flat field* or *plano* lenses.

Staining. Most bacteria are difficult to see with the ordinary bright-field light microscope because there is little contrast between the cell and the surrounding medium. The simplest way of increas-

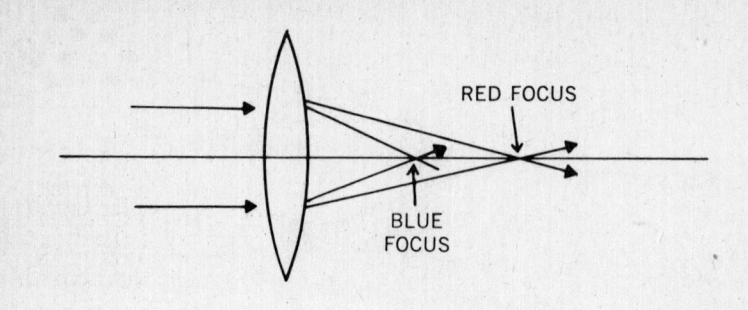

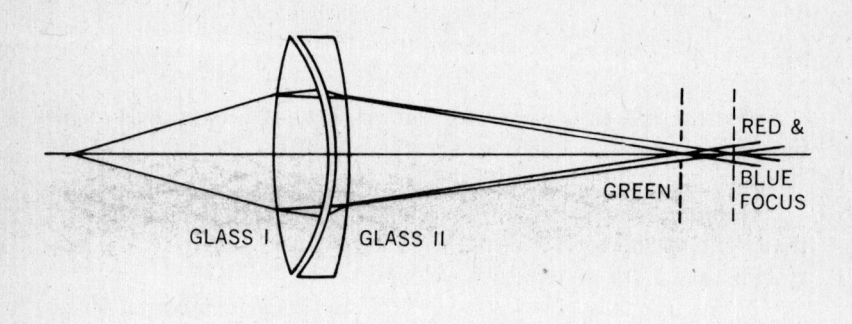

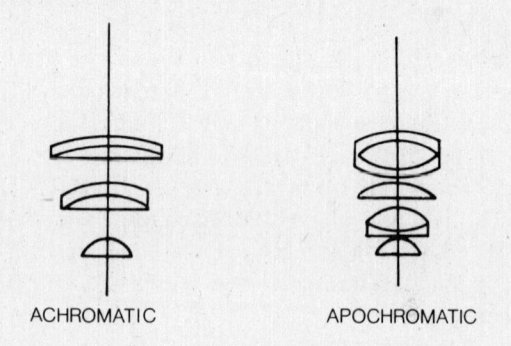

Fig. 2.3. Chromatic aberration.

ing contrast is to dye the cells. A bacterial *smear* is usually heat-fixed on a microscope slide before staining.

<u>Differential Staining.</u> These procedures selectively stain different types of bacteria or particular cellular organelles. The most commonly used differential stain is the *gram stain* named after the Danish bacteriologist who developed it. On the basis of their staining after the procedure, bacteria can be divided into two groups: gram-positive and gram-negative. *Gram-positive* bacteria retain an I_2-crystal violet complex and are purple, whereas *gram-negative* bacteria are stained the color of the counterstain, usually one of the red dyes, such as safranine. Another important differential stain is the *acid-fast stain,* which is used to detect *Mycobacterium* species (tuberculosis, leprosy, etc.) and related actinomycetes. Acid-fast bacteria retain carbolfuchsin even when exposed to nitric acid in alcohol during the decolorizing procedure. Non-acid fast bacteria take up the counterstain, methylene blue. Special stains reveal flagella, bacterial capsules, spores, and a few other cytoplasmic inclusions.

Preparation of Living Specimens. Living organisms can be examined through the light microscope if they are suspended in liquid.

<u>Wet Mount.</u> The specimen in its liquid medium is dropped on a slide and covered with a coverslip.

<u>Hanging Drop.</u> A coverslip containing the specimen in a drop of liquid is inverted over the depression in a special concave slide.

<u>Negative Staining.</u> In this process, the background, not the organism, takes up the stain, which is India ink, or nigrosine. Negative staining is particularly useful in demonstrating bacterial capsules.

DARK-FIELD. A special type of condenser converts a bright-field microscope into a dark-field one (Fig. 2.4). The light from the condenser is directed toward the object in a very wide-angled hollow cone. Only the light which hits the object and is reflected up into the objective forms part of the image. A specimen in dark field appears bright and self-luminous against a black background much like the moon against a dark sky. The dark-field microscope is particularly useful for the examination of living, unstained microorganisms suspended in fluid preparations. It is widely used in the visualization of syphilis spirochetes.

ULTRAVIOLET. Because the wavelength of ultraviolet light is shorter than that of visible light, greater resolution is theoretically

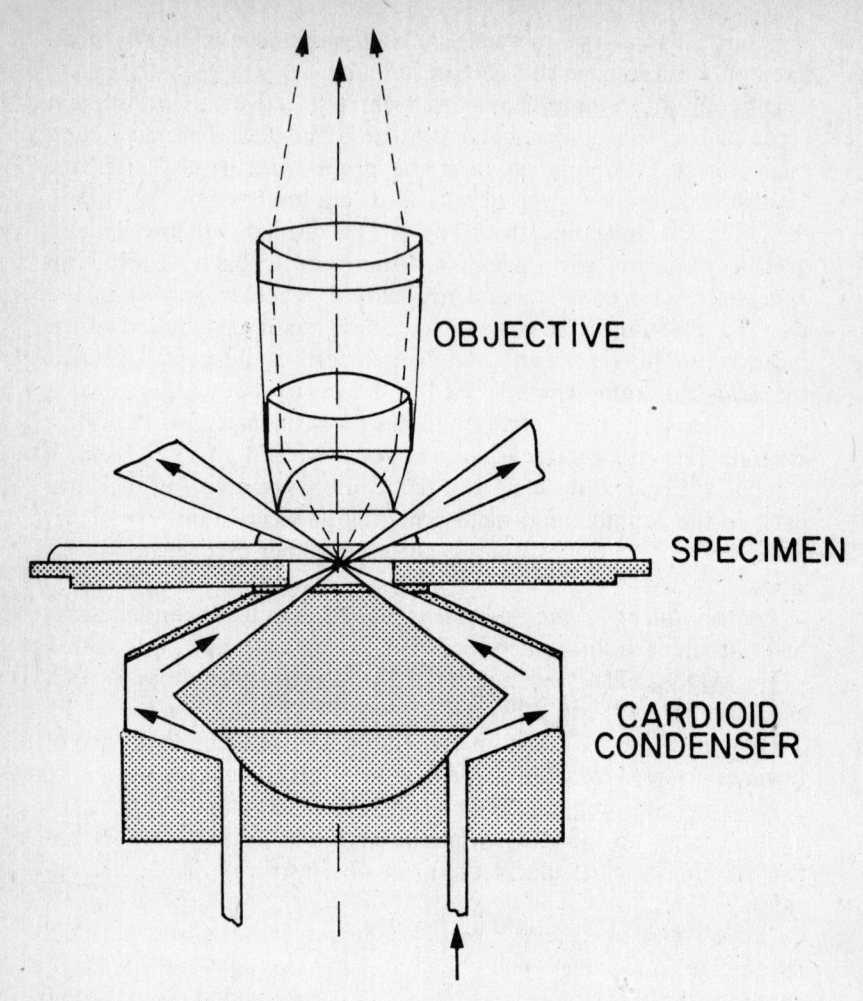

Fig. 2.4. Dark-field microscope.

attainable with an ultraviolet microscope than with a bright-field microscope. Our eyes, however, cannot see ultraviolet light and are injured by it. Objects are viewed instead as images on a fluorescent or a television screen, or on a photographic plate. Ultraviolet microscopy has another disadvantage—it's lethal to living cells. One special, very useful application of ultraviolet microscopy is *fluorescent microscopy*. Specimens are stained with special dyes called fluorochromes which absorb light in the ultraviolet range and fluo-

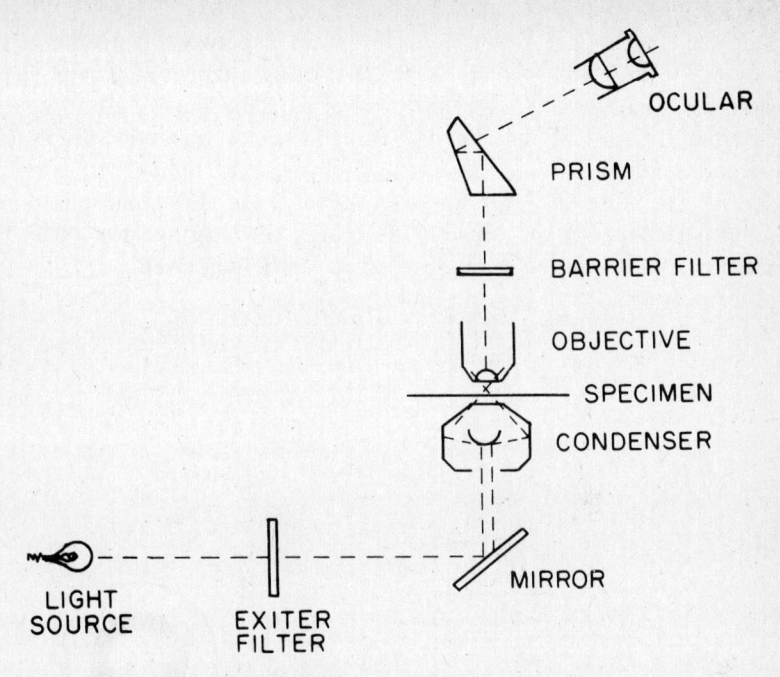

Fig. 2.5. Fluorescence microscope.

resce (give off light) in the visible range. The light pathway in a fluorescence microscope setup is the same as in a dark-field microscope (Fig. 2.5). An exciter filter is placed in front of the light source so that only light with a long wavelength passes through to strike the specimen. Microorganisms stained with fluorochrome dyes or by fluorescein-tagged antibodies are seen as yellow, yellow-green, orange, or red self-luminous cells against a black background. A special ultraviolet-light blocking filter (barrier filter) is placed just below the ocular to protect the eyes. The filter stops the ultraviolet rays but permits the visible light to pass through so that the viewer can see the object. One of the most common uses of the ultraviolet microscope is with the fluorescent antibody technique.

PHASE-CONTRAST MICROSCOPY. The phase-contrast microscope is one of the most valuable tools for studying details in unstained living cells. Small differences in optical path length and density or index of refraction of various structures in a specimen (e.g., nucleus versus cytoplasm) are changed into brightness differences by phase contrast. In order to effect this change, an annular diaphragm is placed in the optical system just below or in between elements of the condenser, and a matching phase-shifting element is placed at the rear focal plane of each objective (Fig. 2.6). The small differences in the optical path length and indices of refraction in the image slow down or shift the phase of some of the light that passes through. Because the eye cannot perceive phase shifts, the phase-shifting element in the objective is designed to change the phase changes into brightness differences that can be perceived.

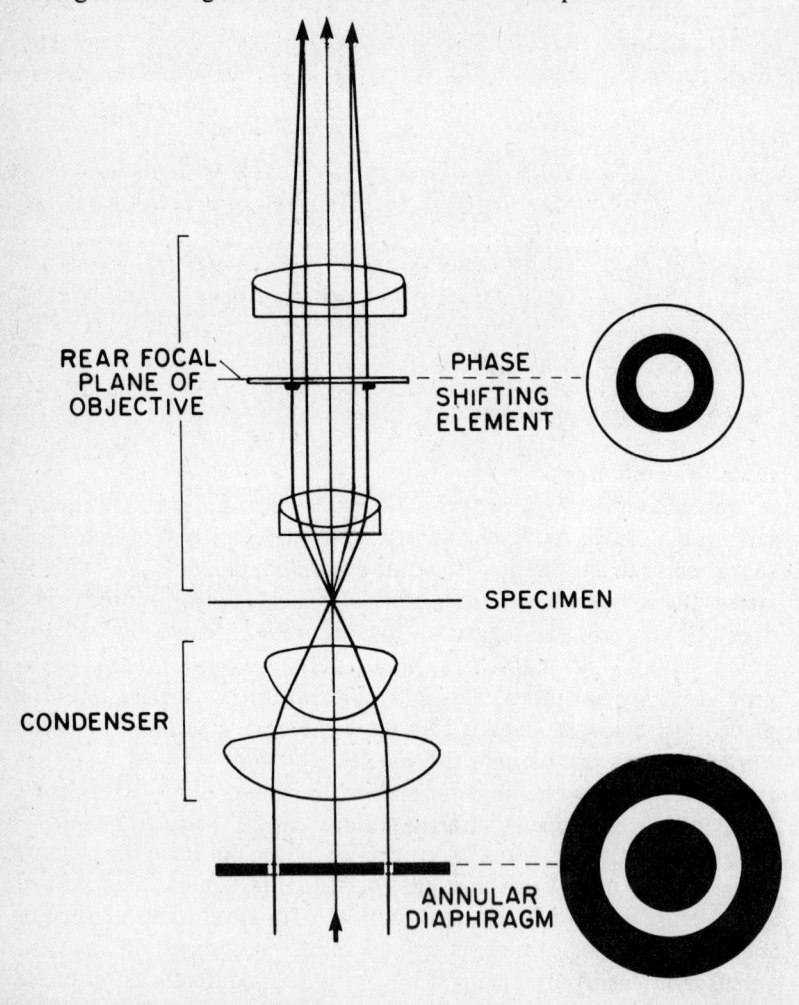

Fig. 2.6. Phase-contrast microscope.

INTERFERENCE MICROSCOPY. Interference is a relatively new form of microscopy which is also used in the study of living cells. Various manufacturers use different systems to obtain interference, but the effects are very similar. An interference microscope converts small differences in optical path length, indices of refraction, and qualities that cause rotations of the light in different parts of the specimen into differences in brightness and color. All the light that enters the interference microscope is polarized by a filter located under the condenser. A crystal plate directly above, or built into the condenser, splits the light into two perpendicularly polarized components (Fig. 2.7). Above the objective and below the ocular is a third special element, a prism or crystal plate which recombines the split beams.

Electron Microscopy. There are two basic types of electron microscopes: the transmission electron microscope (TEM) and the scanning electron microscope (SEM). Both microscopes use electrons instead of light to form images; electromagnets instead of glass lenses to focus the electron beam. Because electrons have a

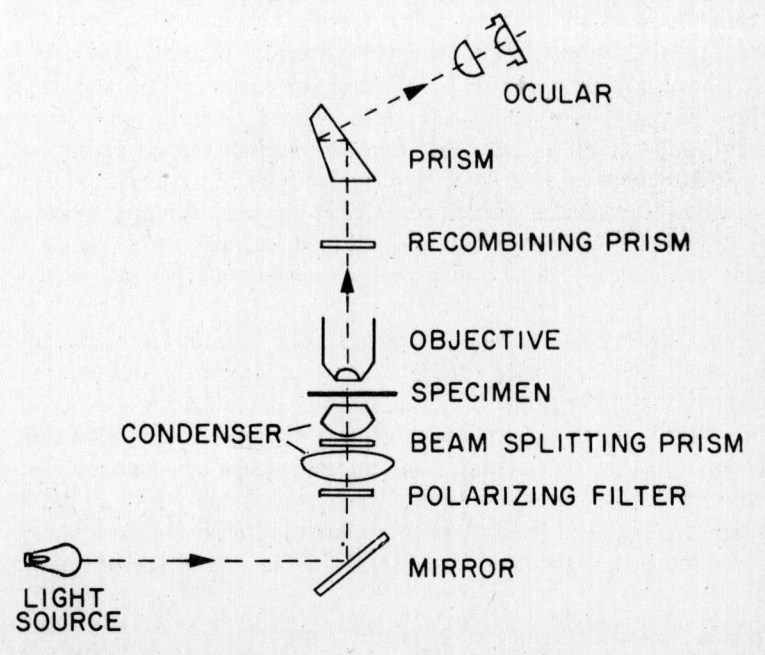

Fig. 2.7. Interference microscope.

very short wavelength (~5 pm),* the resolving power of an electron microscope (TEM) is more than 100× that of the light microscope, and it is possible to usefully magnify images more than 200,000×. Living cells cannot be studied in either type of electron microscope since the column in which the specimens are placed must be evacuated to prevent the electron source (filament) from rapidly burning out and the electron beam from interacting with air molecules. Specimens are usually stained or coated with heavy metals (e.g., gold, lead, and paladium), which absorb or scatter electrons. Images are viewed on an airtight screen or a photographic plate.

TRANSMISSION ELECTRON MICROSCOPE. In general design, the transmission electron microscope (TEM) is similar to an upside-down light microscope (Fig. 2.8). Elaborate preparation of specimens is usually necessary. Cells are chemically fixed, dehydrated, and embedded in plastic. Extremely thin sections of the plastic, cut with diamond or glass knives, are placed on supporting copper screens called grids. These sections may be stained in lead or uranium salts to increase contrast. *Freeze-etching* technique permits the examination of the internal cell components without the potential artifacts that might be produced by chemical fixation. The specimen is rapidly frozen at −100°C and the block fractured with a glass knife, exposing cellular surfaces. The specimen is then freeze-dried and coated with carbon; it is the replica that is examined. Special techniques have been developed for locating chemical constituents of cells. For example, ferritin (iron-containing, electron dense) labeled antibodies can be combined with particular cell antigens so that they can be seen in the transmission electron microscope.

SCANNING ELECTRON MICROSCOPE. The scanning electron microscope (SEM) is used for detailed examination of the external surface of microorganisms. The specimen is placed on a solid flat disk (stub) and then coated with gold or platinum under a vacuum. The stub is then placed in an electron microscope in which the electrons are focused into an extremely sharp *probe* only 5 to 10 nm in diameter (Fig. 2.9). In a manner similar to a television camera, the probe scans the specimen. Because of the fineness of the beam and

---

*A picometer (pm) is one trillionth of a meter (1×10^{-12}m).

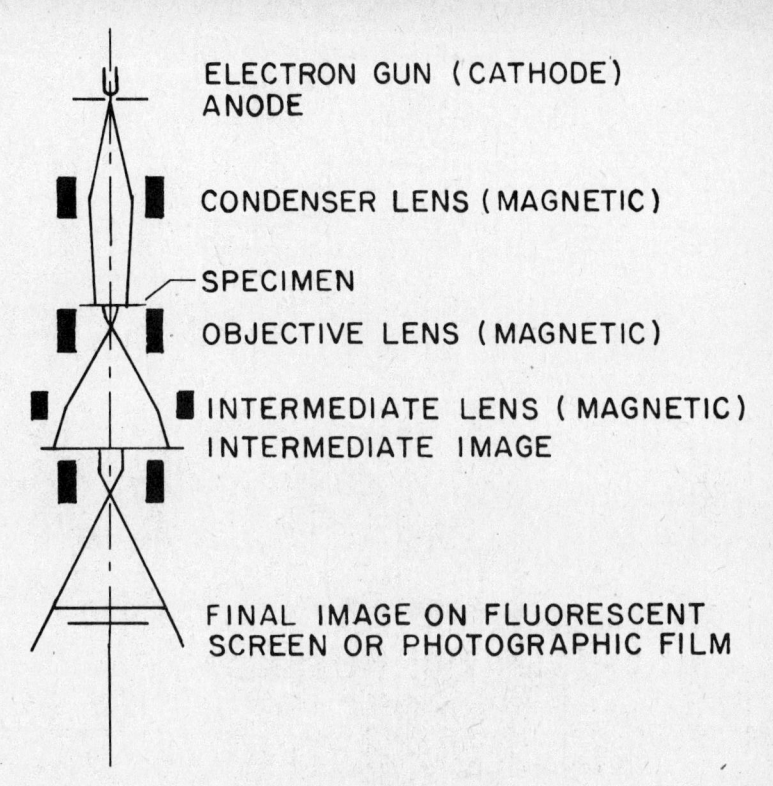

```
            ELECTRON GUN (CATHODE)
            ANODE

            CONDENSER LENS (MAGNETIC)

            SPECIMEN
            OBJECTIVE LENS (MAGNETIC)

            INTERMEDIATE LENS (MAGNETIC)
            INTERMEDIATE IMAGE


            FINAL IMAGE ON FLUORESCENT
            SCREEN OR PHOTOGRAPHIC FILM
```

Fig. 2.8. Transmission electron microscope.

the short wavelength of electrons, the probe can enter very narrow openings and resolve details as small as 10 to 20 nm with good magnifications of up to 100,000×. While the SEM cannot resolve details as small as a TEM can, or magnify a specimen as much, it has a depth of field almost 500× greater than that of a TEM. The image viewed on an SEM screen is formed by secondary electrons backscattered from the specimen and magnetically deflected to a collector or detector. The secondary electron detector is synchronized with the probe beam, and the image seen on the viewing television screen represents successive areas that have been traversed by the probe beam. The scanning time of the scope can be adjusted to allow time for adequate concentrations of electrons to develop an image of the specimen at each point.

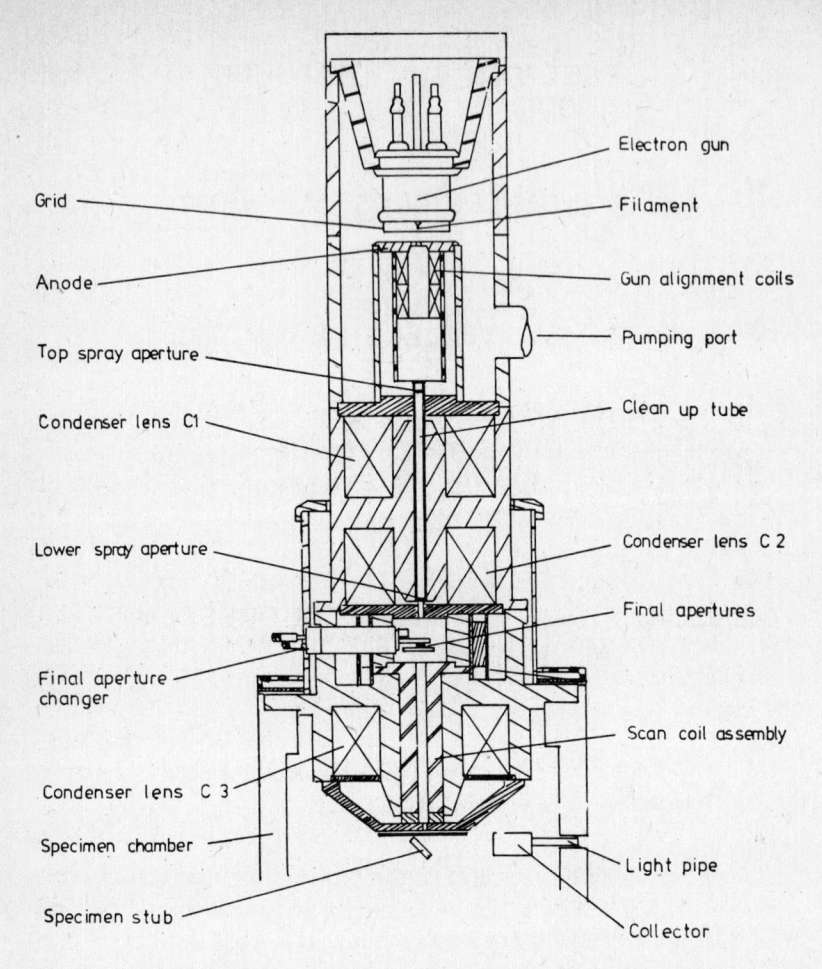

Fig. 2.9. Scanning electron microscope. Courtesy Cambridge Instruments, Inc.

CULTURE METHODS

More information is gathered from studies of populations of microbes than from studies of individual cells. Pure cultures (*axenic*), that is, cultures with only one species, are almost always required for physiological or biochemical studies. The separation of different species from an unknown mixture of organisms (*agnotobiotic*) requires skill and practical knowledge not only of isolation techniques but also of the media and conditions for the growth of the isolates.

Isolation Techniques. The practice of isolation using *sterilized* (the total absence of any form of life) media and instruments is known as *aseptic* (no contaminating microorganisms) technique. There are two general methods for isolation and separation of microbes: (1) surface spreading (streak-plate) methods; and (2) emulsion dilution (pour-plate) methods. Whichever method is used, the separated individual or small clumps of microbes begin to multiply in place as the plate is incubated. Soon isolated populations become large enough to be recognizable as *colonies*. A colony consisting of the descendants of a single cell is known as a *clone*.

SURFACE SPREADING (STREAK-PLATE) METHODS. In surface spreading methods, a drop of the inoculum is transferred to the surface of a medium solidified with agar by means of a sterile inoculating needle or loop, a cotton swab, or a pipette. The microorganisms are then spread or streaked and thus diluted on the surface of the agar by means of a sterile inoculating needle, an inoculating loop, or a glass spreader.

EMULSION DILUTION (POUR-PLATE) METHOD. The emulsion dilution method differs from the surface spreading method in that the agar medium is inoculated while it is still liquid so that individual bacteria can be distributed throughout. In practice, test tubes containing the medium are placed in boiling water until the agar is melted, then transferred to a water bath of about 45°C. The 45°C temperature is warm enough to keep the medium liquid but cool enough not to kill most microorganisms on contact. The medium is removed from the water bath, rapidly inoculated and mixed, and then poured into a sterile petri dish. To ensure well-isolated colonies, serial dilution is carried out in test tube transfers before the medium is poured into the petri plate.

OTHER METHODS. The above methods work well with bacteria, yeasts and many fungi, some protozoa, and some unicellular algae. Viruses may also be inoculated and separated by modifications of these methods. In addition, protozoa, unicellular algae, and some fungi can be isolated from agnotobiotic mixtures by migration techniques, dilution techniques, selective toxicity (antibiotics or antifungal drugs), or by laboriously picking up individual cells with a fine inoculating loop or a fine bore pipette. The picking of individual cells is usually done with the aid of a dissecting microscope or a micromanipulator. If the inoculum presents a hazard (biohazard) to the person making the isolation, the risks are reduced by working in special transfer rooms, hoods, or glove boxes designed for the purpose. The aim of isolation is normally to obtain a pure (axenic) culture. Sometimes this cannot be achieved. The microorganism may be able to grow only in, or in the presence of, another organism(s). The growing together of known species of organisms is known as *synxenic* culture. If only two species are involved, the culture is *monoxenic*. The successful maintenance of two-membered cultures requires considerable skill because a stable biological balance between the two living components is essential.

Types of Media. Because microorganisms differ greatly in their nutritional requirements, no single medium or set of growth conditions will promote the growth of all microorganisms present in a natural population equally well.

COMPLEX (EMPIRICAL) MEDIA. Microbiologists of the last century grew bacteria in complex (empirical) media which were extracts or digests of natural products such as milk, meat, blood, fruits, vegetables, plants, yeast cells, and soil. Traditionally and practically, complex media are still used for the isolation and maintenance of many groups of microorganisms. Nutrient broth (which contains beef extract) is an example of a complex medium. Although the chemical composition of complex media is not exactly known, they contain all of the required elements and a rich assortment of organic compounds that satisfy the requirements of many microbes, and they are generally of sufficiently consistent quality to be reliable. Complex media are generally lower in cost than other media and are useful for the cultivation of organisms with unknown or multiple nutritional requirements and as media for the maintenance of culture collections that include organisms with different nutritional requirements.

SYNTHETIC (ARTIFICIAL, HOLIDIC, CHEMICALLY DEFINED) MEDIA. Synthetic media are formulated of known inorganic and organic compounds. Because they have reproducible composition, they are useful in precise nutritional, physiological, biochemical, and genetical studies. However, they are often expensive and time-consuming to prepare. The exact nutritional requirements of a number of organisms are still incompletely known. Media for such organisms are mainly synthetic, enriched with small amounts or extracts of natural materials. Such media are known as *partially defined* or *oligidic* media.

SELECTIVE MEDIA. The goal of many isolations is the separation of particular species or types of microorganisms from diverse natural communities. Selective media, that is, media designed for such separations, can be divided into two types on the basis of their approach to selection: *enrichment cultures* and *selective inhibition*.

Enrichment Cultures. The design of enrichment-culture media is quite complex and beyond the scope of this brief treatment. One example of a selective enrichment medium is the one used to isolate lactic acid bacteria. These anaerobic bacteria are remarkably resistant to lactic acid which they themselves produce in their fermentation of sugar. The medium for the selective enrichment of lactic acid bacteria contains a large amount of a fermentable sugar, glucose (20 g/l), and yeast extract (10 g/l), a rich source of growth factors. Raw milk, sewage, and rotting vegetables are used as inocula. While many species initially grow in the medium (which is incubated anaerobically), lactic acid, the end product of many fermentative bacteria, accumulates, creating conditions less favorable for most types of bacteria but favorable to lactic acid bacteria.

Selective Inhibition. An example of a selective inhibitory medium is MacConkey's agar, which is used to isolate and distinguish particular gram-negative, potential disease-causing bacteria from enteric samples. MacConkey's agar contains two inhibitory substances, bile salts and crystal violet. The crystal violet in the concentration used inhibits the growth of most gram-positive bacteria and the bile salts inhibit the growth of most nonenteric gram-negative bacteria.

CHARACTERIZATION MEDIA. Certain media are designed for studying the specific or potential metabolic activities of microorganisms in order to characterize and identify them. Many of these media have an indicator substance(s) which demonstrates a change in a particular substrate or the formation of a particular product or class of products. If it is toxic or inhibitory, the indicator substance(s) must be added to the medium after the growth and activity have already occurred. An example of a characterization medium is triple sugar iron agar (TSIA), which contains 1% lactose and sucrose, 0.1% glucose, $FeSO_4$, peptone and phenol red (a pH indicator). The medium is initially red. If after incubation a black precipitate forms, then the bacterium has the ability to metabolize sulfur-containing amino acids in the peptone. The H_2S formed as the result of such activity reacts with the $FeSO_4$ to produce FeS, the black precipitate. If the medium changes from red to yellow, acid products have been produced as the result of fermentation. Bubbles in the agar indicate gas production. The medium turns purple if alkaline end products are produced from the metabolism of the amino acids in the peptone. If glucose is fermented, the medium may first turn yellow as the glucose is fermented and later turn red again as the acid products are aerobically metabolized or if the peptone is digested. The alkaline return is less likely to take place if either lactose or sucrose is fermented since so much acid may be produced that the organisms that produce it are unable to grow in the medium.

STERILIZATION METHODS

The process of sterilization frees the object treated from all living organisms. Sterilization can be achieved by exposure to lethal physical or chemical agents, or, in the case of liquids, by mechanical separation. Not all microorganisms are killed instantly by exposure to the sterilizing agent; some are more resistant than others (Fig. 2.10). Some species of microorganisms form spores that are very resistant to lethal agents. For example, while most bacteria are killed in boiling water after 2 or 3 minutes, spores of some species can withstand boiling for many hours.

Physical Sterilization Methods. There are many practical methods for sterilizing materials and the method of choice often depends upon convenience as well as the nature of the materials to be sterilized.

AUTOCLAVE. The most widely used equipment for sterilization is the *autoclave*. It can be used to sterilize metallic instruments, dressings, glassware, and most, but not all aqueous media. Autoclaves are like pressure cookers. Laboratory autoclaves are commonly operated at a steam pressure of 15 lb/in^2 at a temperature of 120°C. Complete sterilization of most materials in an autoclave is accomplished within 15 to 30 minutes, the variation in time being due to the volume of material sterilized and surface-to-volume ratios.

DRY HEAT. Dry heat is used principally to sterilize glassware or other heat-stable solid materials. The objects are wrapped in paper or otherwise protected from later contamination and then placed into a circulating hot air oven at 170°C for 1.5 to 2 hours.

GASEOUS AUTOCLAVE. The recent wide use of plasticware in disposable syringes, petri dishes, culture tubes, and other equipment

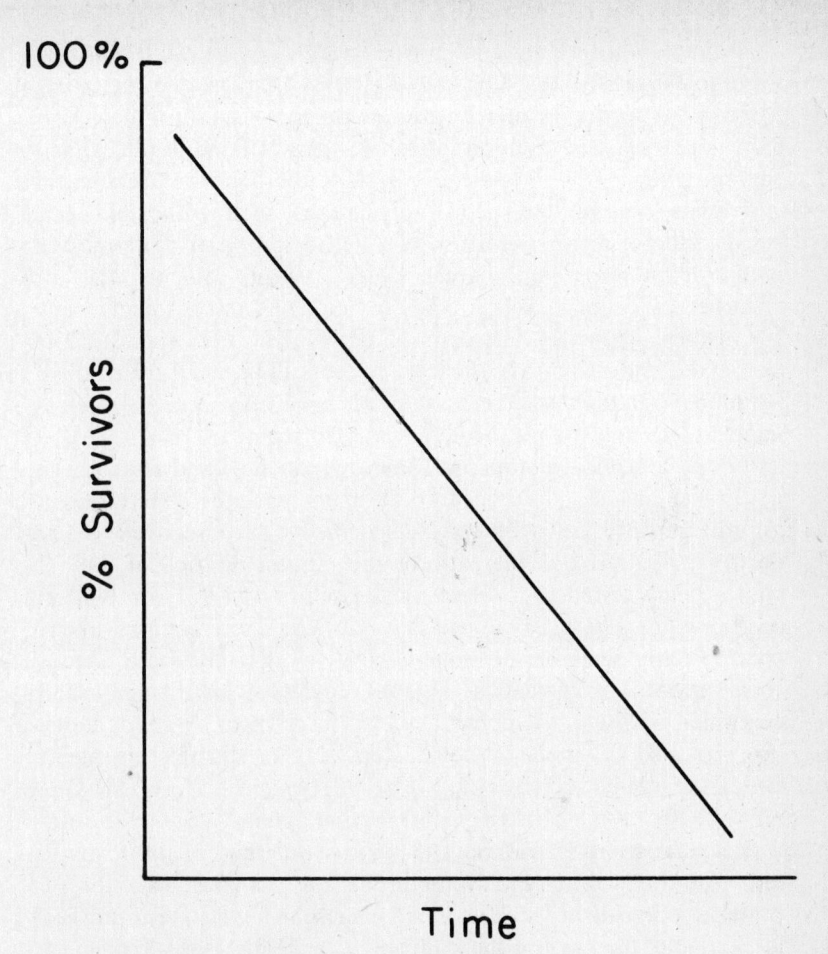

Fig. 2.10. Resistance of microorganisms to sterilization.

and for plastic-backed adhesive bandages has hastened the development of a new type of autoclave for heat-labile materials—the gaseous autoclave. Ethylene oxide is the most commonly used gas in such instruments, but propylene oxide and formaldehyde have also been used. Economic decisions determine the sterilizing regimens in this predominantly commercial process. At 20 to 50 percent humidity and with 720 mg/l of ethylene oxide, sterilization can be achieved at 37.7°C (100°F) in 8 hours or at 54.4°C (130°F) in 4 hours.

RADIATION. Sterilization of the same types of heat-labile materials is sometimes achieved commercially with x rays, and rarely with gamma (γ) waves. In one English commercial establishment, bandages are sterilized by rotation for 3½ days around a ^{60}Co gamma ray-emitting source. Another physical method for sterilization which has some limited special applications, such as the continuous sterilization of air, is the ultraviolet germicidal lamp. The most effective lethal wavelengths are between 250 and 300 nm; they are absorbed by cellular aromatic amino acids and nucleic acids.

Chemical Agents. A distinction must be drawn between sterilization and disinfection. *Sterilization* is the killing of all forms of life while *disinfection* is the removal of all harmful species of microorganisms. Among the most commonly used surface-sterilizing agents and disinfectants are ethyl and isopropyl alcohols and phenol. Phenol is used as a standard with which to compare the effectiveness of all other disinfectants. The *phenol coefficient,* as effectiveness standard, is calculated as the ratio of the highest dilution of the substance being tested to a dilute phenol solution (1:90) that will kill *Salmonella typhi* in 10 minutes but not in 5. There are a number of commercially available phenolic derivatives that are more effective than phenol. Heavy metals (Hg, Ag, and Cu), which are usually combined with organic compounds to make them safer for human use, are also commonly used as sterilizing or disinfecting agents. Oxidizing agents such as H_2O_2, ICl, $NaIO_3$, and $NaClO \cdot 5H_2O$ are also effective as sterilizing or disinfecting agents.

Sterilization by Filtration. The principal method of sterilizing liquids containing heat-sensitive materials such as vitamins and serum proteins is by filtration. Historically microbiologists sterilized heat-labile liquids by passing them through sterile filters made from diatomaceous earth, asbestos fibers, or sintered glass. Since these types of filters are hard to clean and have other disadvantages, they have been largely replaced by newer types of disposable cellulose acetate or polycarbonate membrane filters held in stainless steel or glass funnels.

ENUMERATION METHODS

At the outset a student might ask why there are so many different methods for estimating the biomass and numbers of microorganisms in a sample. Nothing would seem simpler than counting the number of cells directly as seen in a microscope. Experience, however, has shown that not all the microbes one sees in the microscope are alive (viable) or active. Thus each of the many methods for counting (enumerating) microorganisms has advantages and limitations that must be considered before making a choice. Often speed and convenience are the most important criteria.

Cell and Colony Counts. Crowding and clumping severely limit the accuracy of cell-counting methods. In most cases samples must be diluted to a range that is suitable to the method chosen.

DIRECT METHODS. Direct counting of total cells suffers the limitation mentioned above: that not all cells counted in a sample are alive or active.

Microscopic Methods

Counting Chambers. Bacteria and small unicellular algae, yeasts, and very small protozoa are often counted directly under the microscope in special counting chambers. The most commonly used chamber is a hemocytometer, designed for making blood cell counts. There are many styles of hemocytometers, but they are generally microscope slides with a square-millimeter central square of very smooth glass or glassplated with a thin layer of silver that is very finely subdivided into smaller squares. The chamber has edges that hold a coverglass 0.1 mm above the surface grid. The total volume of a hemocytometer is thus 0.1 mm^3. Microorganisms are introduced into the chamber by pipette. Larger chambers, such as the Sedgewick-Rafter chamber, holding 0.1 to 1 ml, have been designed for the counting of plankton, protozoa, and most unicellular algae. Considering the time for setup and washing and drying the chambers between samples, the process of making direct cell counts under a microscope is one of the most time-consuming enumeration methods.

Fluorescence. The counting of bacteria found in soil samples, detritus, and other particulate samples is very difficult. One of the newest techniques for estimating the numbers of bacteria on the surfaces of such samples is to add the dye acridine orange and study them in an epifluorescence microscope. Living bacteria take up the dye and fluoresce ("light up"), making it easy to distinguish them against a black background. An epifluorescence microscope differs from an ordinary fluorescence microscope in that the ultraviolet light is directed onto the specimen from the top surface rather than having to be transmitted through the specimen from below.

Coulter Counter. A Coulter counter is an electronic instrument that automatically counts the number of particles in a known volume of sample. As bacteria or other small organisms pass through a small window (aperture) in the apparatus, they displace an electronic charge between the inside and the outside of the window. The apparatus is designed so that it can not only measure the total number of particles that have passed through the window but also sort them into categories by their total volume. Different windows are used, depending on the size range of microorganisms being counted.

Membrane Filter Counts. Microbial populations in some natural environments are so low that it is sometimes necessary to concentrate them before counting. Large volumes (e.g., 1 liter) are passed through a membrane filter imprinted with a grid pattern. The filter can then be stained and cut into strips that will fit on a microscope slide. When immersion oil is added to the filter-slide preparation, the filter becomes transparent, leaving only the stained bacteria and the grid for examination under an oil immersion lens of a compound microscope.

INDIRECT METHODS. Indirect cell or colony counting methods rely on growing microorganisms from the initial sample. If the initial sample is a mixed population then it is possible that the medium will select and grow only a fraction of the population. This can be used to advantage if particular organisms are the focus of the work.

Dilution Methods. The number of bacteria and other microorganisms present in water, sewage, milk, ice cream, and most food products is usually determined by one of the two methods discussed below. They have the advantage that a large number of samples can be rapidly processed relatively inexpensively. In large laboratories, many aspects of the process can be automated.

<u>Most Probable Number (MPN).</u> The sample is diluted tenfold in a series of successive media. The tubes in higher dilutions are made in duplicate or triplicate. The media is then incubated. Growth will be observed in all the tubes at the lowest dilutions but at some point in the series growth will be observed in some tubes but not in others. Mathematical tables can be consulted which, on the basis of probability, suggest the most probable number of microorganisms in the original sample.

<u>Standard (Viable) Plate Count.</u> The initial sample is also diluted in a tenfold series. Samples are then inoculated into melted media

for emulsion plating or inoculated and spread on the surface of a petri plate. Some plates in the dilution series contain so many colonies that they are impossible to count; others contain only a few colonies. Plates with 30 to 300 colonies are considered optimal for estimating the number of bacteria in the initial sample. There are some new electronic instruments that can automatically count the colonies on a standard size (100-mm) petri dish.

Membrane Filtration Method. If the sample is initially dilute (e.g., some lake, river, or marine samples), it can be filtered through a bacteria-retaining (0.2 to 0.4 μm pore) sterile membrane filter. The filter can then be aseptically removed from the apparatus and placed on the surface of an appropriate medium in a petri plate. The nutrients in the medium are sucked up into the filter by capillary action, and after incubation, colonies form on the surface of the filter. As in the standard plate count method, plates with 30 to 300 colonies are used to estimate the initial number of microbes in the volume filtered.

Cell Mass and Related Methods. These convenient and quick methods give relative measures of the number or mass of microorganisms in a sample. They are often used to corroborate measurements obtained by one of the cell or colony counting methods.

TURBIDOMETRIC METHODS. A cell suspension looks turbid because each cell scatters light. The more cells present, the more turbid the suspension. There are a number of different instruments that can be used for measuring either the scattered light (nephelometers) or the absorbed (colorimeter or spectrophotometer) light (Fig. 2.11). Within a range, optical density is proportional to biomass. Nephelometry is a more sensitive method at low levels of turbidity. To relate turbidity to cell number a calibration curve must first be constructed for the organism in question, since larger organisms scatter more light than smaller ones.

PIGMENT ABSORPTION. Spectrophotometers, or fluorometers, are sometimes used to estimate the biomass of unicellular algae present in a sample. The concentration of pigments per cell varies under different environmental conditions, a factor that must be considered when making biomass estimates.

DRY WEIGHT. The mass of some types of microorganisms, particularly fungal mycelia, are sometimes measured directly. The procedures vary according to the type of organism, but all depend upon washing the cells free of medium and drying them to a constant weight.

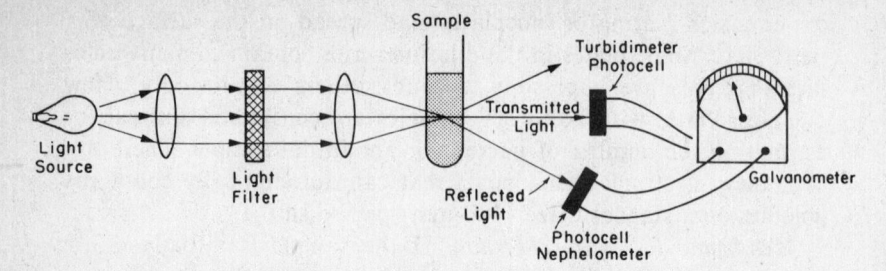

Fig. 2.11. Turbidometric methods.

CHEMICAL METHODS. Sometimes it is useful to estimate microbial biomass by making quantitative determinations of substances present in fairly constant amounts in living cells. The Kjeldahl method measures total nitrogen. Another method estimates protein by means of the folin reagent. A recent procedure uses enzymes extracted from firefly tails to measure microbial ATP or total energy charge in a sample.

Chapter 3

PROCARYOTIC CELL STRUCTURE

Living cells fall into two distinct organizational types: the procaryotic and the eucaryotic. Both have adaptive advantages that have permitted them to coexist for perhaps over a billion years. Of the two cell types, procaryote cells are structurally much less complex. Eucaryotic cells have unit membrane bounded internal subdivisions or organelles that perform specialized physiological and metabolic functions. The only membrane system in the vast majority of procaryotic cells is the cytoplasmic membrane and its infoldings.

GENERAL MORPHOLOGY OF BACTERIA

The procaryotic cell varies considerably in size and form.

Size. Most of the bacteria commonly studied in the laboratory, such as *E. coli, Bacillus, Lactobacillus, Salmonella,* and *Pseudomonas,* are approximately 0.5 to 1 μm in diameter and 2 to 10 μm in length. At either extreme are members of the order Microplasmatales, with very small coccoid forms only 0.1 to 0.3 μm in diameter and members of the *Cytophaga* or *Saprospira* groups (*Cytophaga, Sporocytophaga, Flexibacter, Saprospira,* and *Leucothrix*) which develop filaments that reach 100 to 500 μm or sometimes more in length. Cyanobacteria and related colorless forms attain millimeter to centimeter length.

Form. Bacteria fall into one of four generally recognized forms: *coccus, bacillus, spirillum,* and *branched filamentous.*

COCCI. Cocci (singular, *coccus*) are spherical or ellipsoidal cells which can be found singly or in characteristic arrangements related to the patterns of their division. *Diplococci* are cocci which divide in one plane and remain attached predominantly as pairs. *Streptococci* are also cocci which divide in one plane but remain attached in longer groups or chains. Some cocci characteristically divide in planes that are at successive right angles to each other, producing groups of four cubical packets of cells. *Sarcina* is the best known example of a genus of bacteria with this type of arrangement. Another pattern is found in *Staphylococci,* in which cells divide irregularly in three planes, producing clumps, clusters, or bunches, of cocci.

BACILLI (singular, *bacillus*) are cylindrical or rodlike. Bacilli divide only in one plane so that patterns of arrangements are usually more characteristic of the stage of growth or cultural conditions. Some bacilli, however, are exceptions to this generalization. Cells of *Corynebacterium diphtheriae,* the bacterium that causes diphtheria, tend to produce groupings that look like Chinese characters or palisades. One group of gram-negative bacteria, which includes *Caulobacter* and *Asticcacaulis,* develops extracellular adhesive holdfasts or *prosthecae* at one end of the cell for attachment to the substrate.

SPIRILLA (singular, *spirillum*) are spiral shaped. The length and amplitude of the spiral varies greatly from the short "C" of comma bacteria or vibrios to the long spiral found in the genera *Leptospira, Rhodospirillum,* and *Methanospirillum.*

BRANCHED FILAMENTOUS. These forms are found among gram-positive bacteria belonging to the Euactinomycetes. The long branching filaments are known as *hyphae.* Parts of the mycelium are specialized to produce spores.

SURFACE STRUCTURES

Bacteria vary greatly from group to group in the structures found outside the cell wall. The most common surface structures are flagella, pili, capsules, and prosthecae.

Flagella. Bacterial flagella are thin (~20 nm), hairlike appendages which originate in granular structures (basal body) just below the cell membrane and protrude through the cell wall for distances one to several times the length of the cell.

STRUCTURE. Flagella have three recognizable parts: basal body, hook, and filament. All of these structures are morphologically unlike their analogs in eucaryotes. The basal body consists of a central rod inserted into four sets of rings, the L, P, S, and M rings. The S and M rings are attached to the cell membrane; the other two are anchored in the wall. The filament is composed of subunits of a protein known as *flagellin,* which is rich in the amino acids aspartic and glutamic acids, but which lacks histidine, tryptophan, cysteine, hydroxyproline, and cystine. One unique amino acid, ε-N-methyl lysine, is found in flagellin. The protein is organized in two or three helically wound fibers with amplitude and wavelength characteristic for each species.

ARRANGEMENT. Flagella are arranged differently in various groups of bacteria.

Polar Flagellation. The flagella are attached at one or both ends of the cell.

Lophotrichous Flagellation. A tuft of flagella arises at one or both ends of the cell.

Peritrichous Flagellation. The flagella are not localized but grow from many places on the cell surface.

Pili (Fimbrae). Pili are considerably shorter than flagella and are much more numerous on the surfaces of those cells that have them. Chemically they are similar to flagella. They are not involved in motility. One type of pilus (F or sex pilus), found on male strains of enteric bacteria, has a mating function; it serves as the connection through which DNA is transferred to the recipient. Another type of pilus is involved in the transfer of the resistance transfer factor from donor to recipient cells. Other types of pili enable bacteria to stick to surfaces and to agglutinate (clump together) to form thin films or pellicles on the surfaces of media.

Capsules and Slime Layers. Some bacteria produce a slimy or gummy layer composed of polysaccharides, polypeptides, or polysaccharide-protein complexes, exterior to their cell walls (exopolymers). The ability to produce these layers, and their chemical composition, vary among strains of the same species. In some species of pathogens (e.g., *Diplococcus pneumoniae*) the possession of capsules confers pathogenicity since such strains are not readily engulfed by phagocytes. The distinction between slime layers and capsules is somewhat arbitrary. Slime layers are loose and diffuse, whereas capsules are more compactly bound to cells. It is by means of a slime layer that certain bacteria (e.g., those produced by the organisms that form dental plaque) attach to a substrate.

Prosthecae. These are broad filiform protruberances for attachment that issue from one or more sites on the surface of some gram-negative cells. The prostheca of *Caulobacter* consists of a membranous core surrounded by both the inner and outer cell wall layers.

CELL WALLS

The chemical composition of polymers that form in the cell walls of bacteria is one of their distinguishing features.

Peptidoglycan. Except for some unusual bacteria from extremely saline environments, and one group with no cell walls at all, bacteria have cell walls with peptidoglycan polymers.

CHEMICAL STRUCTURE. Peptidoglycan polymers are composed of two acetylated amino sugars, N-acetylglucosamine and the closely related N-acetylmuramic acid, and a small number of unusual D-amino acids (Fig. 3.1). The N-acetylglucosamine and the N-acetylmuramic acid form chains of from 10 to 65 disaccharides joined in a β (1,4) linkage. The carboxyl group of the N-acetylmuramic acid provides a site for the attachment of a short peptide chain, the most common chain being a tetrapeptide of L-alanine, D-glutamic acid, *meso*-diaminopimelic acid, and D-alanine (Fig. 3.2). Adjacent peptides are cross-linked by peptide bonds between the free amino group of the diaminopimelic acid on one chain and the terminal D-alanine on an adjacent chain. There are many variations in peptide chains among different types of bacteria. Commonly other diamino acids replace diaminopimelic acid and are important in cross-linking the chains. In spirochetes, L-ornithine replaces *meso*-diaminopimelic acid. There is evidence that the peptidoglycan layer is one continuous, extremely large, bag-shaped molecule which completely encloses the cell.

FUNCTION OF THE PEPTIDOGLYCAN LAYER. The peptidoglycan layer confers rigidity and shape to bacterial cells. Bacterial cells have such high internal solute concentrations in relation to their outside environment that, if the cell wall were not present, enough water would enter to cause them to burst. This idea can be verified experimentally. After treatment with *lysozyme*, an enzyme from tears and hens' eggs which hydrolytically cleaves N-acetylmuramyl–N-acetalglucosamine linkages in glycan strands, bacterial cells swell and then burst. The lethal effect of penicillins is similar. These antibiotics inhibit the terminal step in peptidoglycan synthesis, the cross-linking of peptide chains. The resultant weakening leads to osmotic lysis. If the treated cells are placed in an isosmotic

Cell Walls

```
    ─── N-acetylmuramic acid         N-acetylmuramic acid ───
       N-acetylglucosamine          N-acetylglucosamine
       N-acetylmuramic acid  ─── N-acetylmuramic acid
       N-acetylglucosamine          N-acetylglucosamine
    ─── N-acetylmuramic acid         N-acetylmuramic acid ───
       N-acetylglucosamine          N-acetylglucosamine
       N-acetylmuramic acid  ─── N-acetylmuramic acid
```

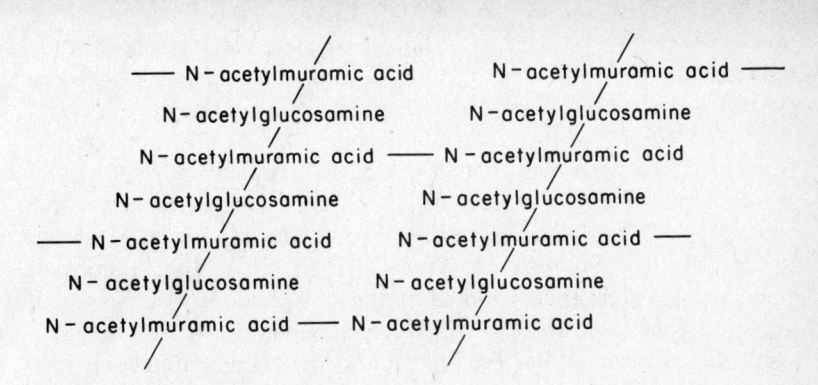

Fig. 3.1. Peptidoglycan polymer.

medium, commonly a sucrose medium, they become spherical but will not burst. In these osmotically balanced media, cells completely stripped of cell wall components are known as *protoplasts;* those with some fragments of the cell wall are usually called *spheroplasts.*

Teichoic Acids. The peptidoglycan matrix of the wall is covalently linked to other macromolecular constituents known as teichoic

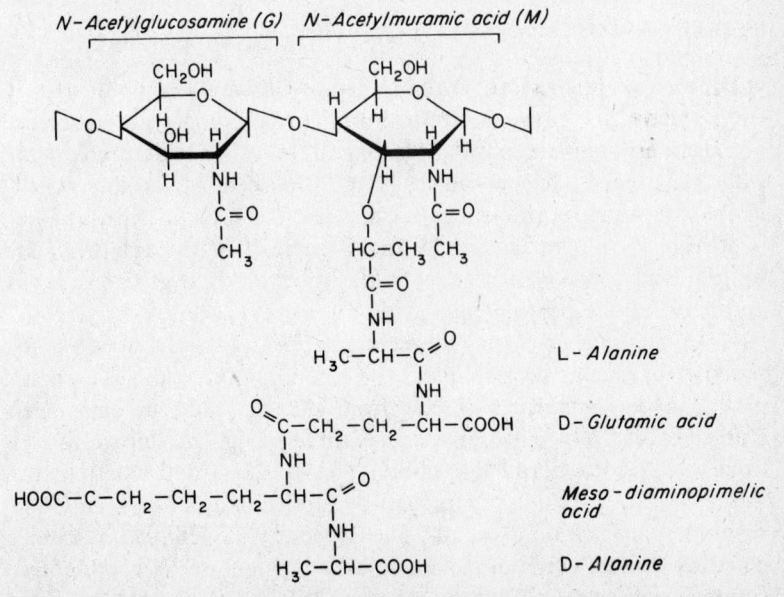

Fig. 3.2. Structure of one of the repeating units in the peptidoglycan cell wall of bacteria.

acids which are water-soluble polyphosphate polymers containing either ribitol or glycerol residues joined through phosphodiester linkages (Fig. 3.3). Chains of 10 to 30 repeat units have glycerol joined by 1,3 or 1,2 linkages, ribitol joined through 1,5 linkages, or have more complex bonds with either sugar residues or N-acetylglucosamine. Most teichoic acids contain D-alanine attached to the second position of glycerol, to the third or fourth position of ribitol, or one of the sugar residues. Much of the teichoic acid fraction of the cell wall is covalently linked to muramic acid. However, a small amount, consisting entirely of glycerol teichoic acids, is associated with the cell membrane. Teichoic acids are also major surface antigens in some gram-positive species.

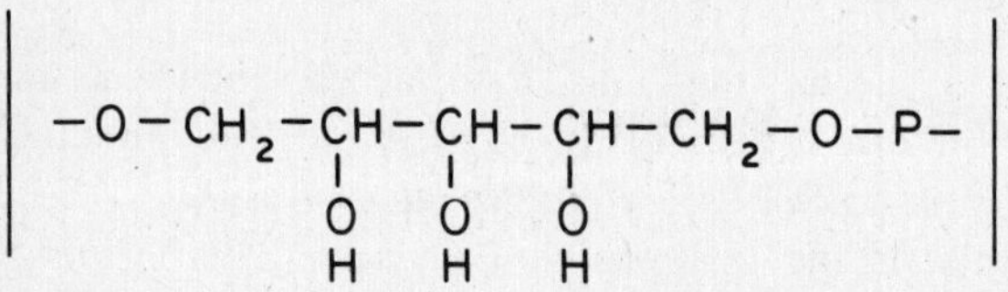

$$-O-CH_2-CH-CH-CH-CH_2-O-P-$$
$$\,|\,|\,|$$
$$OOO$$
$$HHH$$

Teichoic acid backbone
(polyribitol phosphate)

Fig. 3.3. Teichoic acid backbone (polyribitol phosphate).

Differences in the Cell Walls of Gram-Positive and Gram-Negative Bacteria. In gram-positive bacteria the peptidoglycan is a thick, homogeneous matrix covalently linked to other macromolecular wall constituents. It constitutes 40 to 90 percent of the dry weight of the cell wall. In gram-negative species, the layer of this polymer is so thin it seldom exceeds 5 to 10 percent of the cell wall dry weight, and there is less cross-linking between the glycan strands and between the peptide chains.

Gram-negative bacteria, however, have an outer-wall layer composed of proteins, phospholipids, and lipopolysaccharides. The complex lipopolysaccharides (LPS) from enteric bacteria have been studied extensively. The polymer has three recognizable regions: (1) lipid A; (2) core; (3) O side chain. The lipid A fraction is unusual in that the fatty acids are attached by ester linkages to N-acetylglucosamine instead of glycerol. The R core of *Salmonella* species contains a short chain of sugars (glucose, galactose, N-acetylglucosamine) and two unusual substances, 2-keto-3-deoxyoctonoic acid

(KDO) and heptose. The often branched O side chain of four or five sugars contains many common hexoses (six-carbon sugars) such as galactose, glucose, rhamnose, and mannose, as well as one or more very unusual di-deoxy sugars: abequose, colitose, paratose, or tyrelose.

The lipopolysaccharides of gram-negative bacteria are toxic to animals and cause fever, internal hemorrhages, and shock when introduced into the blood stream. They are also major surface antigenic determinants (see Chapter 12).

Cell Wall Structure and the Gram Stain. The reaction of cells to the gram staining technique is not directly linked to bacterial cell wall chemistry but to physical characteristics related to wall thickness. This is why the totally unrelated yeasts (eucaryotic organisms) stain gram-positive.

PLASMA MEMBRANE

The plasma membrane of procaryotes is intimately associated with the cell wall. As in all living cells, its main components are phospholipids and proteins and its principal function is that of a selective semipermeable barrier. The plasma membrane is a typical *unit membrane* built of two protein layers with a lipid layer in between.

Osmotic Barrier. There are some differences, however, between the lipids in the plasma membranes of eucaryotes and procaryotes. Firstly, while sterols are significant components of the plasma membranes of eucaryotes, they are abundant only in the plasma membranes of the *mycoplasma* group of procaryotes. Secondly, some of the fatty acids of eucaryotic plasma membranes are polyunsaturated (more than one double bond) while among the procaryotes plasma membranes contain only saturated or monounsaturated fatty acids. Blue-green bacteria are the exception. They are able to synthesize polyunsaturated fatty acids.

Electron Transport—Respiratory Function. In procaryotes the membrane with its intracellular intrusions is also the site of enzyme systems involved with electron transport and oxidative phosphorylation, functions served by the mitochondria of eucaryotic cells.

Active Transport. The membrane also has enzymes called *permeases* which selectively transport certain metabolites into cells.

Cell Division. In some species the bacterial chromosome is closely associated with sites on the cell membrane. Chromosome replication and new cell wall formation are initiated at these sites.

Other Metabolic Activities. In the photosynthetic bacteria the plasma membrane is extensively infolded and often forms platelike lamellae which bud into vesicles. Photosynthetic activities are associated with these membranous infoldings. Extensive internal membrane systems are also common in the nitrifying and methane-oxidizing bacteria. In many groups the membrane infoldings are localized into lamellar structures in concentric whorls known as *mesosomes* or *chondriods*. There is a wide variation in their appearance in different groups.

CYTOPLASMIC STRUCTURES

A number of different structures are found in the cytoplasm of bacteria.

Ribosomes. The cytoplasm of bacteria is filled with *ribosomes,* the active sites of protein synthesis. The ribosomes of bacteria are small spherical particles composed of RNA and protein, with a molecular weight of about 3×10^6 and a sedimentation constant of 70S. They are smaller than the ribosomes of eucaryotes, which have molecular weights of approximately 1×10^7 and sedimentation constants of 80S. During translation, ribosomes are linearly aggregated into temporary associations known as polysomes (polyribosomes, ergosomes).

Granular Reserves. Different procaryotes store a variety of reserve materials which are visible in light or transmission electron microscopy as granular cytoplasmic inclusions.

GLYCOGEN OR POLYGLUCAN. Blue-green bacteria, many species of enteric bacteria, *Bacillus,* and *Clostridium* store glycogen or starch.

POLY-β-HYDROXYBUTYRIC ACID. Other species of *Bacillus,* enteric bacteria, *Pseudomonas, Rhizobium, Spirillum,* and *Sphacrotilus* store poly-β-hydroxybutyric acid.

CYANOPHYCIN. Many of the blue-green bacteria accumulate a copolymer of the amino acids arginine and aspartic acid known as cyanophycin.

VOLUTIN OR METACHROMATIC GRANULES. Many microorganisms have granules that stain red with a basic dye methylene blue. As linear polymers of orthophosphate, these granules store inorganic phosphate.

SULFUR. Some sulfur bacteria, the purple sulfur and nonphotosynthetic sulfur bacteria, such as *Beggiatoa* and *Thiothrix*, temporarily store inorganic sulfur as cytoplasmic inclusions.

POLYHEDRAL BODIES. The cytoplasm of many blue-green bacteria commonly contains polyhedral bodies; their function is unknown.

Gas Vacuoles (Vesicles). Many aquatic procaryotes, including some purple sulfur bacteria, blue-green bacteria, *Halobacterium, Thiothrix, Pelonema, Microcyclus,* and a few other genera, have gas-filled vesicles which keep them floating. Gas vacuoles enable their possessors to regulate their buoyancy in order to gain a position in the water column where the light, oxygen concentration, nutrient concentration, or other requirements are optimal for metabolic activity. The vesicle membrane is composed of protein subunits arranged to confer the required rigidity.

Chlorobium Vesicles. In the green bacteria, photosynthetic pigments are enclosed in cigar-shaped, membrane-bound vesicles arranged in a cortical layer which underlies and is budded from the cell membrane.

Cyanobacterial Thylakoids. An extensive system of flattened, branching vesicles called thylakoids is found in the outer regions of the cells of blue-green bacteria. In some genera they are arranged in concentric shells at the edge of the cells; in others they are more branched and penetrate most of the cytoplasm. They are the only procaryotic cytoplasmic organelles that have a unit membrane construction. Thylakoids contain the photosynthetic pigments and the enzymes associated with photosynthesis.

Carboxysomes. Some purple bacteria, blue-green bacteria, nitrifying bacteria, and thiobacilli have cytoplasmic structures with polygonal profiles called carboxysomes (polyhedral bodies). They are surrounded by a monolayer membrane and contain the enzyme ribulose diphosphate carboxylase (carboxydismutase), a key enzyme associated with CO_2 fixation.

NUCLEOID

The DNA of procaryotes is a long, continuous, double strand which is folded back on itself many times. Though the DNA is not separated from the rest of the cell by a membrane as it is in encaryotes, the nucleoid, the region where the DNA is concentrated, can be seen in a transmission electron microscope.

Because the bacterial chromosome is naked DNA (not associated with basic proteins as is the DNA of eucaryotic chromosomes), some microbiologists prefer to distinguish it by calling it a *genophore*. At a specific point on its circumference, the bacterial chromosome is associated with a *replicator site* on the cell membrane (*mesosome*). Replication begins with the formation of a new replicator site adjacent to the old one. One strand of the DNA is broken and one end (the 5' end) attaches to the new replicator site. The chromosome rotates past the attachment site that has the replication enzymes. The daughter chromosomes are separated as the sites on the membrane spread farther apart during growth and cell elongation.

PLASMIDS

Many bacteria contain plasmids, small circular molecules of DNA (mol wt $\sim \times 10^7$; Col E, mol wt 4.2×10^6, R477, mol wt 1.7×10^8) that carry genes for their own replication. Different plasmids carry genes that govern conjugation (F factors), confer resistance to drugs (R factors), and transfer information for biochemical pathways or the production of toxins. Plasmids able to integrate with host chromosomes are called *episomes*. Genetic exchanges between similar plasmids are common.

Because of their medical significance, R factors have been extensively studied. R factors have two distinct subunits: resistance transfer factor (RTF) and r-determinants (resistance determinants). The RTFs carry the genes for replication and transmission of the plasmid. The inhibitory substances for which resistance has been found include antibiotics, sulfonamides, and some heavy metals (mercury, cadmium, zinc, arsenic, nickel, and cobalt). The resistance factor for penicillin is a gene that codes for a potent enzyme, penicillinase, which destroys the antibiotic by hydrolyzing the β-lactam ring. The plasmid factor that mediates chloramphenicol resistance is an enzyme that acetylates the antibiotic. Genes carried on plasmids code for the synthesis of the proteinaceous toxins (bacteriocins) liberated by strains of bacteria active against closely related strains. The toxins liberated by *E. coli* are *colicins* (e.g., Col B, Col E, Col I, and Col V).

Plasmids become integrated into the host chromosome at *insertion sites* (insertion elements) that contain specific sequences of ~ 1000 nucleotides. An enzyme, integrase, coded for by either the host's chromosome, a plasmid, or a phage, recognizes the insertion sequence and integrates plasmids and phages into the host's chromosomes. One group of plasmids, *transposons,* sandwiches a series of genes between insertion elements at either end. Transposons can insert themselves at many sites on a chromosome, bringing their genes along with them. Not all transposons fall into the plasmid category.

The presence of one type of plasmid in a cell may bar the maintenance of a second kind (*incompatible*). *Compatibility* is governed by plasmid genes. In *E. coli* at least 20 compatibility groups have been recognized.

ENDOSPORES

Endospore formation in eubacteria occurs in the genera *Bacillus* and *Clostridium.* It is not a form of reproduction since it involves a reorganization of a vegetative cell into a single specialized cell which is highly resistant to drying, heat, freezing, radiation, and chemical inactivation. Spores have 20 to 30 percent less water than vegetative cells and they have high amounts of calcium (as much as 12 percent dry weight) and dipicolinic acid (5 to 10 percent dry weight). There are also quantitative and qualitative differences in the enzyme complements of spores and vegetative cells. The spore itself is an unusual multilayered structure consisting of two coats, a cortex, and a core.

OUTER SPORE COAT. This is a proteinaceous layer with a high content of cysteine and hydrophobic amino acids. It constitutes about 40 to 60 percent of the dry weight of the spore.

INNER SPORE COAT. The inner coat is a membrane enclosing the cortex.

CORTEX. The cortex is composed of a distinct peptidoglycan.

CORE. The core contains a core wall (the equivalent of the usual cell wall), a plasma membrane, some enzymes and ribosomes, and the nucleoid. Calcium and an unusal constituent, dipicolinic acid, are associated with the core.

PROCARYOTES AND THE ORIGIN OF CERTAIN EUCARYOTE ORGANELLES

Procaryotes share some striking structural and metabolic properties with the *mitochondria* and *chloroplasts* of eucaryotes. All three forms—procaryotes, mitochondria, and chloroplasts—contain DNA in the form of a long circular molecule which is not complexed with histones. They seem to be able to synthesize their own ribosomes (70S), DNA and RNA nucleotide polymerases, and many metabolic enzymes (oxidases, cytochromes, etc). Enzyme systems associated with the plasma membrane and its infoldings in procaryotes are the same as those associated with the membranous infoldings, tubules, or lamellae from the inner membranes of mitochondria and chloroplasts. This evidence has lent support to a theory that these organelles originated as endosymbiotic procaryotes in the early evolution of eucaryotic cells.

Chapter 4

MICROBIAL GROWTH

Growth may be defined as an increase in the quantity of chemical components and cellular structures. As a rule, growth is an increase in the size and weight of individual cells, followed by cell division (binary fission). Microbiologists are primarily concerned with increases in the number of cells, that is, increases in cell populations.

GROWTH RATE CONSTANT AND GENERATION TIME

Two of the most useful quantitative ways of looking at microbial population are the growth rate constant and generation time. The *growth rate constant* is defined as the number of population doublings per unit time, and its reciprocal, *generation time* or *doubling time*, is defined as the time required for the population to double in number. Doubling times can vary greatly depending on the species of bacteria and the conditions required for their growth and reproduction. Under optimum conditions for its growth, one of the most rapidly reproducing bacterial species, *Bacillus stearothermophilus*, has a generation time of 8.4 minutes. Also under optimum conditions, the common intestinal bacterium, *Escherichia coli*, has a generation time of 20 minutes, and the population of the relatively slow growing tuberculosis bacterium, *Mycobacterium tuberculosis*, doubles in approximately 6 hours. Some deep-sea bacteria have even longer generation times of weeks to months.

The numbers of cells in microbial populations are so large that scientific notation (exponents of 10) is used to express them. Thus, a very dense bacterial culture containing 10,000,000 cells/ml is noted simply as 10^7 cells. Scientific notation is used to calculate the growth rate constant k. Because the numbers of cells involved are so large, logarithms are used to simplify the calculation.

$$k = \frac{\log_{10}X_t - \log_{10}X_0}{0.301t}$$

where
- X_0 = number of cells at initial time (inoculum)
- X_t = number of cells at a later time
- t = length of time from X_0 to X_t expressed in minutes or hours
- k = growth rate constant, expressed as number of doublings per unit time (N is another commonly used notational symbol for the growth rate constant)
- $1/k$ = generation time, the time required for the population to double (g is also used to denote generation time)

Example:

$$X_0 = 1000 \text{ cells}$$
$$X_t = 1{,}000{,}000 \text{ cells}$$
$$t = 4 \text{ hr}$$
$$\log_{10} \text{ of } 1000 = 3$$
$$\log_{10} \text{ of } 1{,}000{,}000 = 6$$

$$k = \frac{6-3}{(0.301)\,4\text{ hr}} = \frac{3}{1.204 \text{ hr}}$$

k = 2.49 doublings/hr

$1/k$ = 0.40 hr (24 min), time required for population to double

The measurement of growth is discussed in Chapter 2 under the heading Enumeration Methods.

THE GROWTH CYCLE OF POPULATIONS

Unless very special conditions are met, the growth of a microbial population that is contained ("batch" culture) goes through a number of phases from the time of its establishment to its eventual senescence and death (Fig. 4.1). The characteristic pattern of growth is commonly called the growth curve.

Lag Phase. When transferred to fresh media, microorganisms do not immediately begin to grow and divide, even if the media and all other conditions are favorable for rapid growth. They usually go into a period of adjustment and "tooling up" for balanced growth and reproduction. At this time, regulatory mechanisms within the cells adjust the metabolic machinery of the cell to utilize materials in the fresh media for energy and biosynthesis. This is reflected in the high rates of RNA and protein synthesis observed in this phase. Individual cells are usually much larger than those in other phases in the growth cycle.

The length of the lag phase depends on many factors, all of which relate to how much adjustment in the cell metabolism must be made and the conditions under which it occurs. For example, if cells from a culture in the log phase of growth are transferred to a fresh batch of the same medium incubated under identical conditions, the lag phase will be short or skipped and the cells will continue to grow. If, however, cells in a stationary phase of growth are transferred to the same medium, there will be a lag. The same is true for cells transferred from a very rich, complete medium to a synthetic, minimal medium because many new synthetic enzymes are required for growth.

At the end of the lag phase, cells begin to reproduce, slowly at first, but as time progresses, at an increasing rate. This aspect of late lag is sometimes referred to as the *positive acceleration phase*.

Log (Logarithmic or Exponential) Phase. During this phase the population doubles at regular intervals. The growth rate is maximal for the particular medium and incubation conditions. Cells are physiologically young and biologically active. Biochemical and physiological properties commonly used for the identification of or-

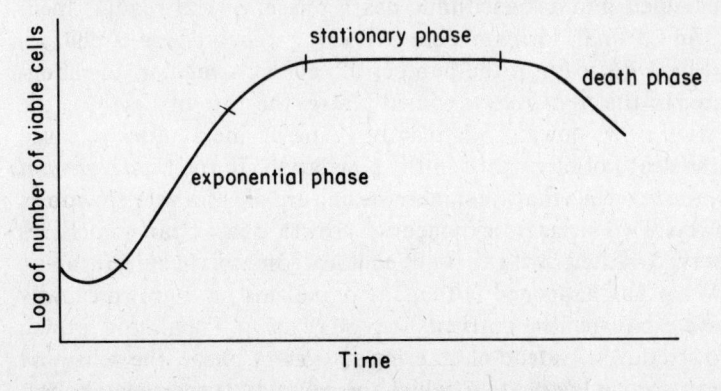

Fig. 4.1. Generalized growth curve for a bacterial population.

ganisms are usually most readily demonstrated during log growth. The average size of the cells is at its minimum for the species. Cell walls are thinnest and metabolic activities at their highest rates.

Eventually the growth of the population changes the condition of the culture. Among many things, food begins to run out, cells become crowded, inhibitory waste products accumulate, pH changes, hydrogen acceptors are used up, and exchange and diffusion rates change. As a result, the rate of reproduction slows down. This portion of the growth curve is often separated from the log phase as the *negative growth acceleration phase,* the period in which the population is increasing at a decreasing rate.

Stationary Phase. At some point in the history of a contained ("batch") culture the number of cells being produced is balanced by the number of dying cells. During this phase cells change physiologically. In some species cell walls thicken. Some cells form spores, others store reserves of various materials. Cells in stationary phase are more resistant to adverse physical (heat, cold, radiation) and chemical agents than those in log phase. A number of factors have been shown to influence the length of stationary phase. In some species, for instance, the stationary phase is longer if the culture is grown 10°C lower than the optimum temperature.

Death Phase; Final Phases. A microbial culture goes into its death phase when the number of viable (living) cells decreases exponentially, the inverse of population growth during the log phase. Some cell populations decline rapidly in a matter of days, whereas others may persist for weeks, months, or years. Where there are recognizable stages in the decline of a culture, the death phase can be subdivided into a logarithmic death phase, a final readjustment phase, and a final dormant phase. The *log death phase* is characterized by a decrease in the number of cells at a regular, unchanging rate. In the *final readjustment phase,* the rate of death of the population slows down. Cells may be living on metabolites released from the death of other cells in the population. In the *final dormant phase,* the few cells that remain in the culture die at a very slow rate.

Diauxy. Two separate exponential growth phases can sometimes be observed in the growth curve of contained microbial cultures (Fig. 4.2). What has happened is that the organisms have grown rapidly and have exhausted the nutrients they were using. Population growth has slowed down. Instead of entering stationary phase, the cells have entered a second lag phase in which the cells adjust their metabolism to take advantage of other nutrients in the medium. A common example of this phenomenon is the growth of many organisms in phenol red glucose broth. Commonly called alkaline return, the broth first turns yellow, indicating acid end-product production as the glucose is used as an energy source, then slowly returns to red as beef protein is metabolized and the alkaline end products neutralize the acid.

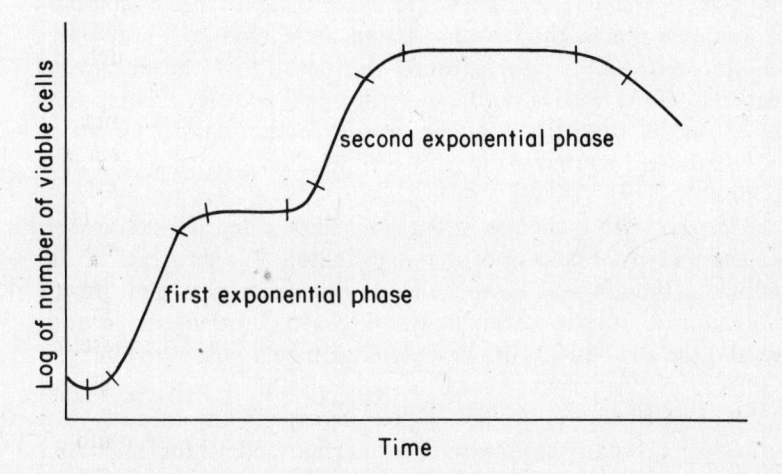

Fig. 4.2. Diauxy. A microbial growth curve with two exponential growth phases.

MODIFICATION OF GROWTH PHASES

Microbiologists have devised ways of manipulating the growth of cultures in the laboratory.

Synchronous Growth. Experimental designs sometimes require cultures in which all cells are at the same stage of the growth cycle. Measurements made on such cultures are in lieu of, or roughly equivalent to (if it were technically feasible to do so), measurements made on single cells. Synchronously dividing cultures of microorganisms can be obtained by a number of techniques, the most common being cyclical manipulation of environmental conditions. For example, synchronous cell division can be induced by repetitive, precisely timed, shifts of cultures from low temperatures where cell division is retarded to temperatures where cell division is rapid.

Continuous Cultures. For many studies it is desirable to keep cultures in continuous exponential growth. This can be attained by using a series of interconnected bottles or flasks. The culture is placed in one flask, the *reaction flask*. Media from a second flask, the *reservoir,* flow into the reaction flask at a rate that keeps the culture in exponential growth. An overflow device in the reaction flask keeps the culture at constant volume and permits the removal of excess media and organisms. Once the system is in equilibrium, cell number and nutrient status remain constant, and the system is said to be in *steady state.* There are two principal types of continuous culture systems: the turbidostat and the chemostat.

TURBIDOSTAT. In a turbidostat, the density of the culture is measured photoelectrically; the electric signal controls a pump that delivers media from the reservoir to the reaction flask. The device can be adjusted so that a range of cell densities can be maintained indefinitely.

CHEMOSTAT. In a chemostat both population density and growth rate are under the control of the investigator. The flow rate of the medium is usually set to a particular value and the rate of growth of the culture adjusts to this flow rate. The conditions for incubation and the constituents of the medium can be adjusted to alter the growth rate. Chemostats have been quite useful in genetic, physiological, and ecological studies with microorganisms. In ecological studies, chemostats have served as useful models for microbial growth in nature under conditions of low nutrient concentration (Fig. 4.3).

FACTORS THAT AFFECT GROWTH RATES

Nutrient Concentration. In many respects, bacterial growth is usefully compared to simple chemical reactions. All other things being equal, just as the velocity of a chemical reaction is determined by the concentrations of the reactants, the rate of growth of a microorganism is proportional to the concentration of a limiting nutrient (Fig. 4.4). Curves relating the growth rate to nutrient concentration are typically hyperbolic and fit the equation

$$\mu = \mu_{max} \frac{C}{K_s + C}$$

where

μ = the specific growth rate at limiting nutrient concentration

μ_{max} = the specific growth rate at saturating concentrations of the nutrient

C = initial concentration of the limiting nutrient (sometimes S is used as a symbol for this variable)

K_s = a constant equal to the nutrient concentration supporting a growth rate equal to ½ μ_{max}. It is analogous to the Michaelis-Menten constant of enzyme kinetics.

Temperature. Temperature is one of the most important factors that affect the growth and survival of microorganisms. There is a minimum temperature below which growth no longer occurs, and a maximum temperature (upper thermal limit) above which growth no longer occurs. In the range between the two the growth rate increases rapidly as temperature increases. This is because the chemical and enzymatic reactions in the cells are speeded up. At the upper thermal limit, critical cellular components become irreversibly denatured. Every microorganism has an optimum temperature at which generation time is shortest. The optimum temperature is always much closer to the upper thermal limit than to the minimum temperature.

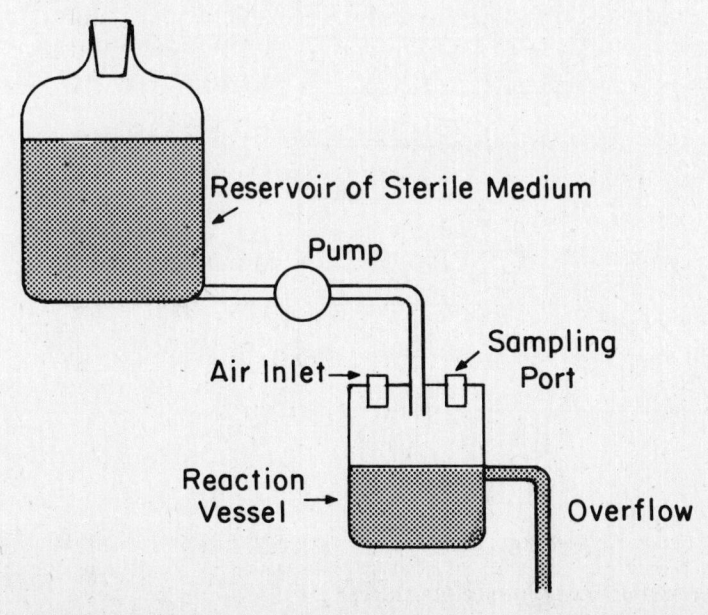

Fig. 4.3. Simplified diagram of a chemostat.

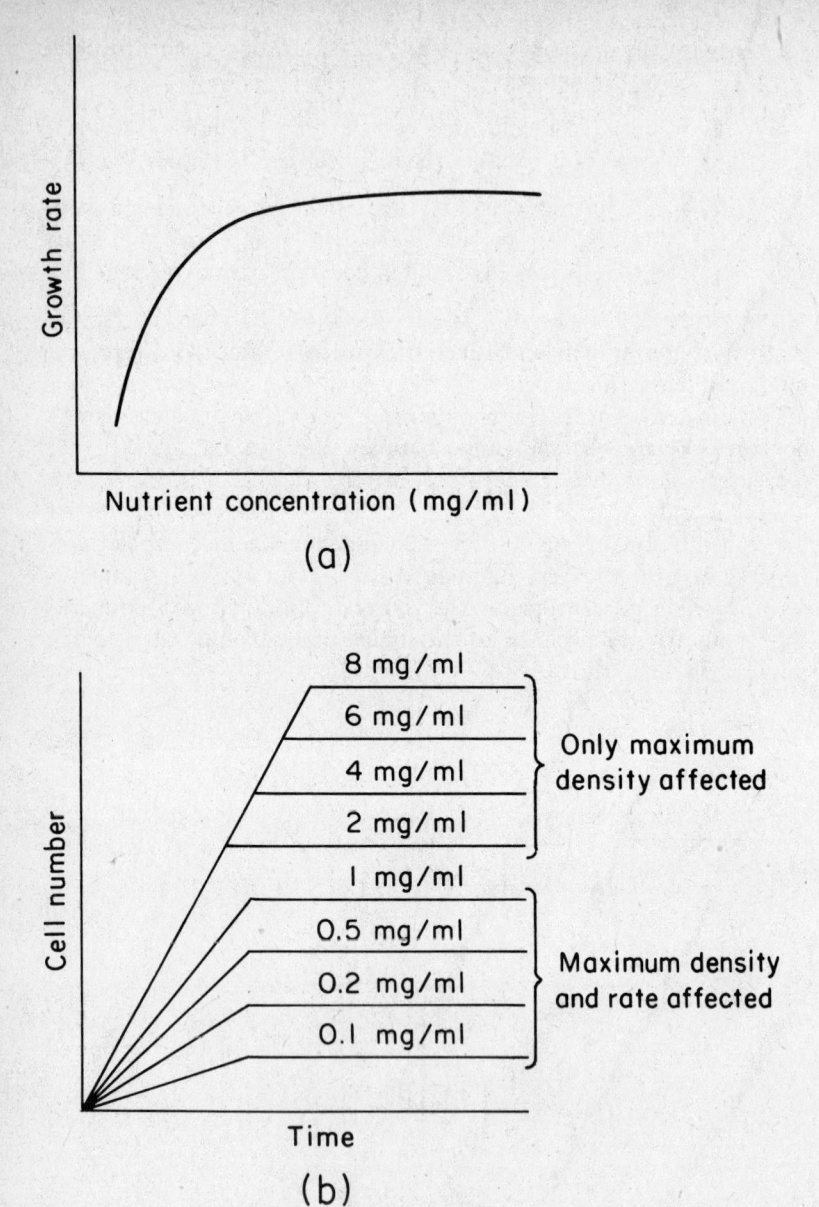

Fig. 4.4. Relationship between nutrient concentration, growth rate, and maximum population density. (a) The effect of nutrient concentration on growth rate. (b) Growth curves at different nutrient concentrations.

Microorganisms have been found to grow at temperatures below 0°C at the bottom of the sea and under polar ice to 75 to 80°C in hot springs. No single species grows throughout this range. The usual thermal range of most microorganisms is about 30 to 40°C. Species with very narrow temperature ranges are called *stenothermal;* those with wide ranges, *eurythermal.*

PSYCHROPHILES (CRYOPHILES). A psychrophile is a microorganism that can grow at 0°C. *Obligate psychrophiles* have temperature optima of 15°C or lower and temperature maxima at or below 20°C.

Although *facultative psychrophiles* can grow at 0°C they usually have optima above 25°C (25 to 30°C) and maxima (rarely) of 35°C or higher. Psychrophiles are common in the oceans, deep lakes, and the snow; however, they are rare in the environments where we harvest and store food. We take advantage of this by using cold or freezing to retard or prevent microbial spoilage of food.

MESOPHILES. Microoroganisms with thermal optima in the range of 25 to 40°C are called *mesophiles*. The minimum temperature of mesophiles is about 10°C, below which even slow growth does not occur. Mesophiles are common inhabitants of the skin and digestive tracts of animals, of forest soils, and every imaginable terrestrial and shallow aquatic habitat.

THERMOPHILES. Organisms that grow at temperatures above 45 to 50°C are called *thermophiles*. Temperatures as high as this are found in nature, in soils, in sand in full sunlight, in fermenting compost piles, and in hot springs. Eucaryotic microorganisms are not found above 60°C. Photosynthetic procaryotes are limited below 73°C. Only a few types of bacteria grow at temperatures above 75°C.

Osmotic Pressure. Microorganisms vary widely in their ability to grow in media with high or low osmotic pressure. A few species of microorganisms are *osmophilic,* that is, they require high osmotic pressures for growth. Other species are *osmotolerant,* that is, they are able to withstand high osmotic pressures. But most microbes do not grow well in media with very high osmotic pressure. When we think of the Dead Sea, we are reminded of the fact that very few organisms can grow there. Long before there were microbiologists to explain the reasons, many people dried fruit, fish, eggs, vegetables, or salted meat to prevent spoilage. There are some special terms that relate to saline or marine environments. Many of these organisms are red.

OBLIGATE HALOPHILES. An obligate halophile is a microorganism that requires NaCl for growth. An *extreme halophile* will grow only when the NaCl concentration approaches saturation. The cell walls of *Halobacterium,* an extreme halophile, differ considerably from those of other bacteria. Its enzymes require high salt concentrations for activity.

FACULTATIVE HALOPHILES. This type of microorganism will grow in relatively high NaCl concentrations, but does not require it.

Hydrostatic Pressure. Some organisms that live on the bottom of the sea seem to require hydrostatic pressure for their growth and reproduction. There are technical difficulties in studying these organisms, so it is not certain if they are barophilic or barotolerant.

pH. Each microorganism has a pH range within which growth is possible, and usually a well-defined optimum pH. Very few microbial species live at a pH less than 2 or greater than 10. Most yeasts and fungi grow better in slightly acid media (pH 5 to 6) and cyanobacteria grow best under slightly alkaline conditions (pH 7.5 to 8.5). The few species of microorganisms that grow well at low pH are usually obligate acidophiles. Because there are not too many acid-tolerant species, acid is often used to prevent the growth of microorganisms in perishable foods. Pickles, sauerkraut, buttermilk, and many fruits and meats stored in vinegar are examples.

Oxidation Reduction Potential. The relative amount of oxygen or its lack affects the growth of many microbial species. Such species are characterized accordingly.

OBLIGATE AEROBES. Species that require the presence of O_2 for growth and reproduction.

FACULTATIVE AEROBES. Species that grow better in the presence of oxygen but do not absolutely require it for growth.

MICROAEROPHILES. Species that require oxygen but at very low levels (less than 0.2 atm).

AEROTOLERANT ANAEROBES. Species that do not require oxygen for growth and do not grow better in its presence.

OBLIGATE ANAEROBES (Aerophobes). Species that do not use oxygen and will not grow in its presence. Obligate anaerobiosis exists in only two groups of microorganisms, bacteria and protozoa. These organisms have been found in sediments and intestinal tracts and are often responsible for the spoilage of canned foods and certain diseases such as tetanus and gangrene.

Chapter 5

MICROBIAL NUTRITION AND METABOLISM

Microbial metabolism is concerned with how microbes produce energy and how they use it.

ENERGY GENERATION

What is the essence of life? Could we recognize a foreign life form? Though these questions seem to be highly philosophical or in the realm of science fiction, the microbiologists who designed the Viking space probe had to consider questions like these. Arguments on the essence of life sooner or later focus on the second law of thermodynamics which, simply stated, is that no process involving an energy transformation will occur unless there is a degradation of energy from a concentrated form to a dispersed form. Or, put another way, things in the universe tend to go from a more organized state to a less organized one. In order to maintain the same state or increase organization in an isolated system, energy must be put into it. Living organisms are such isolated systems. They temporarily stave off dispersion of their molecules by constantly taking in, transforming, consuming, and dissipating energy. Through natural selection, various groups of microorganisms on earth have exploited almost every potential energy-yielding reaction. There are microorganisms that find the energy needed for life in mine tailings, swamp gas, and a host of other seemingly unlikely sources. To understand the circumstances in which microorganisms live, grow, and compete with each other, we must know their nutritional and physiological requirements.

Nutritional Classification of Microorganisms. In classifying microbes according to mode of nutrition, there are two main considerations: their source of energy and their source of carbon.

SOURCE OF ENERGY. As to source of energy, microorganisms can be divided into three groups: phototrophs, chemotrophs, and paratrophs.

Phototrophs. These are organisms that use photochemical reactions for energy generation and storage. Phototrophs can be further divided into two groups, photolithotrophs and photoorganotrophs, on the basis of the source of hydrogen used in the photochemical reaction. In *photolithotrophs,* an inorganic compound is the hydrogen donor; in *photoorganotrophs,* organic compounds donate hydrogen.

Chemotrophs. Chemotrophs are organisms that derive their energy from chemical reactions. This category also can be divided into two subgroups, chemolithotrophs and chemoorganotrophs. In *chemolithotrophs,* inorganic reactions serve as the prime energy source; in *chemoorganotrophs,* the energy source is organic reactions.

Paratrophs. Paratrophs gain all their energy from the host cells they infect.

SOURCE OF CARBON. On the basis of their source of carbon, microbes can be divided into two broad categories: autotrophs and heterotrophs.

Autotrophs. Autotrophy literally means self-feeding. In the context of microbial nutrition, autotrophs are organisms that use CO_2 as their source of carbon.

Heterotrophs. Heterotrophs need organic sources of carbon.

Facultative Autotrophs. Some autotrophs can also obtain carbon from organic sources as do heterotrophs. Such autotrophs are referred to as *facultative autotrophs;* autotrophs that do not have this ability are called *obligate autotrophs.*

Energy generation and storage. In all forms of life, the generation of ATP (adenosine triphosphate) and related compounds is a fundamental mechanism for the trapping of some free energy released by various chemical and photochemical reactions, its storage, and its coupling to biosynthesis. ATP has two particularly reactive high-energy bonds that it is able to donate to a number of metabolic intermediates, thereby converting them to activated forms capable of participating in certain biosynthetic reactions that require a

Energy Generation

$$\begin{array}{c} CH_2OH \\ | \\ HC-O\,\textcircled{P} \\ | \\ COOH \end{array} \xrightarrow{H_2O} \begin{array}{c} CH_2 \\ || \\ C-O\sim\textcircled{P} \\ | \\ COOH \end{array} \xrightarrow{ADPATP} \begin{array}{c} CH_3 \\ | \\ C=O \\ | \\ COOH \end{array}$$

2 Phosphoglyceric Acid — Phosphoenol Pyruvic Acid — Pyruvic Acid

Fig. 5.1. Substrate level phosphorylation in the Embden-Meyerhof pathway.

higher energy level than those metabolic intermediates had before. In cellular metabolism, ATP is generated by two distinct biochemical mechanisms: substrate level phosphorylation and electron transport.

SUBSTRATE LEVEL PHOSPHORYLATION. In substrate level phosphorylation, ATP is formed from ADP by the transfer of a high-energy phosphate group from an activated intermediate of a catabolic pathway. Several of the terminal steps in the Embden-Meyerhof (glycolytic) pathway serve as examples of this phenomenon (Fig. 5.1).

ELECTRON TRANSPORT. This mechanism of ATP generation involves the transfer of electrons between certain carrier molecules with fixed orientation in membranes. One good example is the respiratory chain (cytochrome system) where in certain sequential oxidation steps sufficient energy is liberated to permit synthesis of ATP (Fig. 5.2).

As seen in Figures 5.1 and 5.2, the enzymes (dehydrogenases) that remove electrons and hydrogen ions from reduced substrates have prosthetic groups of coenzymes (NAD+, NADP+, flavoproteins, ubiquinones), which carry $2H^+$ and $2e^-$, but the cytochromes transfer only electrons. Interestingly, several B vitamins, niacin (nicotinic acid), and riboflavin form building blocks of NAD (nicotinamide adenine dinucleotide), NADP (nicotinamide dinucleotide phosphate), FAD (flavin adenine dinucleotide), and FMN (flavin

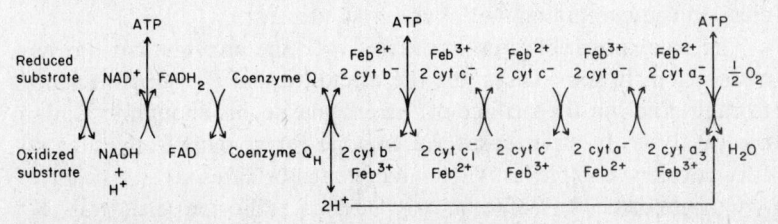

Fig. 5.2. The respiratory chain (cytochrome system).

mononucleotide). In bacteria, the respiratory chain is associated with the cell membrane while in eucaryotes it is associated with mitochondrial cristae.

Enzymes. Many chemical reactions take place spontaneously, but often the rates of reaction are slow because few of the reactant molecules are in an activated state. Chemical reaction rates can be speeded up in several ways, one of which is heating. If the reactant molecules are heated up, more will be in an activated state and the reaction will be hastened.

Because temperatures compatible with life are relatively low, biological systems as they evolved could not take advantage of temperature rise as a means of speeding up their reactions. This function is performed by enzymes, a class of catalysts, which lower the energy of activation. All enzymes are proteins built of long polypeptide chains which are folded three-dimensionally in very specific configurations. The specific conformational structure is essential to enzyme function; if environmental conditions (temperature, pH, osmolarity) are changed enough to unfold the protein, enzymatic activity is lowered or lost. The specific folding leads to the creation of a region on the surface of the enzyme, the *active site,* which is involved in its reactive and catalytic properties. The *substrate* (the compound undergoing reaction) combines temporarily with the active site. After the reaction occurs, the product is released from the surface of the enzyme.

$$\text{enzyme} + \text{substrate} \rightarrow \text{enzyme-substrate complex} \rightarrow \text{product} + \text{enzyme}$$

Some active sites have nearly absolute specificity for a given substrate and do not permit binding by even closely related molecules. Others are far less specific and allow the enzyme to act on an entire class of molecules. If an enzymatic reaction involves two substrates, the second substrate has a combining site on the enzyme that is close to the combining site of the first substrate.

The mechanism of catalysis varies with the enzyme and the type of reaction. In some cases just the orientation of the substrates next to each other on the surface of the enzyme seems enough to explain the catalysis. In other cases the enzyme forms more active, temporary covalent complexes with one of the substrates.

COFACTORS. The cofactor may be a metallic ion (i.e., Na^+, K^+, Mg^{2+}, Ca^{2+} Zn^+) or a relatively complex organic molecule derived from a B vitamin (i.e., riboflavin, thiamine, nicotinamide, pantothenic acid) that serves as a *coenzyme*. The degree to which cofactors are bound to the protein varies. When the cofactor is tightly bound to the enzyme, the nonprotein portion is known as a *prosthetic group* and the protein part is known as an *apoenzyme*. An example of such an enzyme is one of the cytochrome molecules in the respiratory chain in which Fe containing heme is the prosthetic group. Cofactors serve as bridging groups, that is, they bind substrates and enzymes together through the formation of coordination complexes, or they themselves may serve as catalytic centers. An example of the latter is peroxidase, a relatively common microbial enzyme.

ENZYME NOMENCLATURE. Except for a few historically important enzymes (e.g. trypsin, pepsin, and catalase), enzymes are named according to rules developed by an international commission. The names are chemically informative and give the nature of the chemical reaction catalyzed by the enzyme (e.g., hydrolase, transferase, isomerase) and the substrates involved in the reaction. All the names of newly named enzymes end in the suffix -ase. The international names are sometimes so long that scientists often use older names, or give enzymes short nicknames (trivial names) for use in daily conversation (e.g., hexokinase for ATP:hexose phosphotransferase, the enzyme that catalyzes the reaction between ATP and glucose in which a phosphate group is transferred to glucose, yielding glucose-6-phosphate and ADP).

Some reactions can be catalyzed by more than one enzyme. The enzymes are usually distributed in different parts of the cell, function in or are controlled by, different conditions, or are found in different cell types. Because they have the same function, such enzymes have the same name. The general term for two or more enzymes that catalyze the same reaction is *isoenzymes,* or *isozymes*.

MEASUREMENT OF ENZYME ACTIVITY. Enzyme activity is usually determined (assayed) by measuring the rate of disappearance of the substrate, the rate of product appearance or release, or changes in coenzymes. Spectrophotometric techniques are widely used. For example, the activity of an enzyme that uses NAD as a cofactor can be measured in a spectrophotometer. $NADH_2$ absorbs light at 340 nm, whereas NAD does not. In the respiratory chain example given at the beginning of this chapter the rate of substrate oxidation by an extracted appropriate enzyme could be determined by measuring the rate of appearance of $NADH_2$ in a spectrophotometer.

In addition to spectrometric means, enzyme activity is routinely measured (assayed) by respiratory (manometric or polarographic), radioactive tracer, chromatographic, fluorometric, and titration techniques. There are a few other techniques that have less general application, such as microbiological assay.

Within certain bounds, the velocity of enzyme-catalyzed reactions is proportional to enzyme concentration when substrate is in excess and to substrate concentration when enzyme concentration is held constant. As mentioned earlier in this chapter, various environmental factors influence enzyme structure and rate of activity. Some enzymes have very broad tolerances for optimum activity; others do not.

As with most chemical reactions, the rate of enzyme-catalyzed reactions increases with temperature. However, in the case of enzymes there is an upper thermal limit to catalysis beyond which rapid thermal denaturation of the protein causes a loss of activity.

BASIC PATTERNS OF ENERGY TRAPPING AND PATHWAYS OF ENERGY RELEASE

Phototrophy. Among microorganisms, several groups of procaryotes and all the eucaryotic algal groups are phototrophs. The phototrophic procaryotes are the *cyanobacteria* (Cyanophyta, formerly called the blue-green algae), the *green sulfur bacteria* (the Chlorobiaceae, former name Chlorobacteraceae), the *purple sulfur bacteria* (the Chromatiaceae, former name Thiorhodaceae), the *green gliding bacteria* (the Chloroflexaceae), and the *purple nonsulfur bacteria* (the Rhodospirillaceae, former name Athiorhodaceae). In many respects, the photosynthetic processes of the cyanobacteria are closer to those of the eucaryotic algae than they are to those of the other photosynthetic bacterial groups.

COMPONENTS AND MECHANISMS OF THE PHOTOSYNTHETIC APPARATUS. Photosynthetic energy conversion is a complex process that varies in detail between the various procaryotes, the algae, and the higher plants. The photosynthetic machinery in all phototrophic organisms, however, has three main components: light-harvesting pigments, a photosynthetic reaction center, and an electron transport chain.

Light-Harvesting Pigments. Photosynthetic bacteria possess several bacteriochlorophylls, a + b and a + c. Bacteriochlorophylls differ in structure and in absorption spectra from the chlorophylls of algae and higher plants; they absorb light in the near-infrared region (660 to 870 nm). Photosynthetic bacteria also have carotenoid pigments that absorb light between 450 and 550 nm. These carotenoids are responsible for their overall appearances of purple, red, violet, pink, orange-brown, or brown. Cyanobacteria have only one form of chlorophyll, chlorophyll a, and have characteristic water-soluble accessory pigments called *phycobilins*. One class of phycobilins, the phycocyanins, absorb light maximally between 570 and 580 nm. Cyanobacteria with this pigment are red or orange. In the simple cyanobacteria, the photosynthetic pigments are found in concentric lamellae at the periphery of the cell; more complex types have multilayered membrane systems. In the purple photosynthetic bacteria, the photosynthetic pigments are part of an elaborate lamellar or tubular membrane system connected to the plasma membrane. The photosynthetic vesicles of the green photosynthetic bacteria are not continuous with the plasma membrane.

Photosynthetic Reaction Center and Electron Transport Chain. Chlorophyll molecules at the reaction center absorb light energy gathered by the pigments and each loses an electron. These electrons then pass to an electron transport system. The process is called *photophosphorylation*.

Cyclic Photophosphorylation. Bacteriochlorophyll and the initial electron transport molecules are very closely associated with each other. When a bacteriochlorophyll molecule absorbs a quantum of light, the molecule becomes excited and an electron is driven off (Fig. 5.3). In cyclic photophosphorylation, the electron is accepted by the first member of an electron transport chain, a nonheme iron-containing molecule, ferredoxin. At the end of the chain, a new high-energy bond is generated (ATP from ADP) and the electron is returned at a lower energy state to the bacteriochlorophyll at the beginning of the cycle.

Noncyclic Photophosphorylation. In algae and higher plants, photophosphorylation differs from that of photosynthetic bacteria in being noncyclic. There are two different types of photochemical centers. Photosystem I pigments include chlorophyll a and P700, and, in common with bacteriochlorophylls, are most responsive to far-red light. Photosystem II pigments include chlorophyll a_2 and are more responsive to blue light. The key difference between noncyclic and cyclic photosystems is that in the former, the electron ejected from photosystem II after the pigment system is excited by

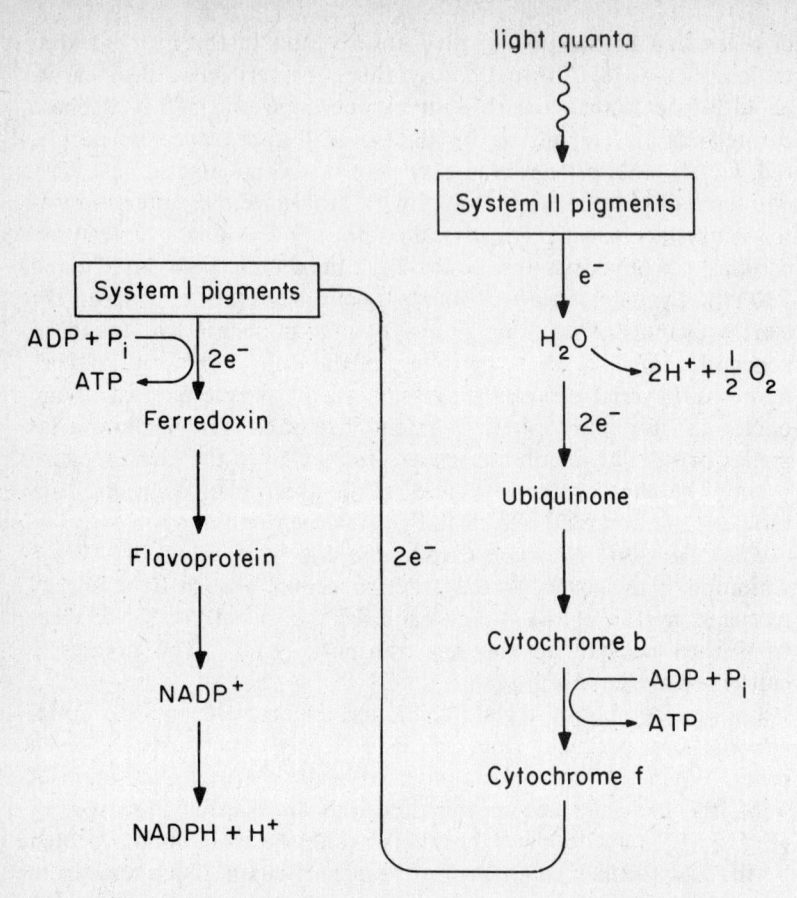

Fig. 5.3. Photosynthetic pathway in cyanobacteria, algae, and higher plants.

a quantum of light, is used in the photolysis of water (Fig. 5.3). With accompanying liberation of O_2, the electron is also passed through a respiratory chain to generate ATP. The electron is not recycled to chlorophyll as it is in cyclic photophosphorylation, but is further routed to reduce $NADP^+$ with the production of another molecule of ATP.

CO₂ ASSIMILATION. Photophosphorylation provides the energy and the chemical reductant for the fixation (reduction) of CO_2. The overall process can be expressed as:

$$CO_2 + 2H_2D \rightarrow (CH_2O)_x + D$$

In the above, the H_2D stands for the hydrogen donor and the

$(CH_2O)_x$ represents a generalized carbohydrate and not any actual compound. In higher plants, algae and cyanobacteria, the H_2D is, of course, water, and oxygen is liberated as an end product.

$$2H_2O + CO_2 \rightarrow (CH_2O)_x + O_2 + H_2O$$

As noted in the discussion of cyclic photophosphorylation, water is not lysed and reducing power is not provided by hydrogen released in the photolysis of water. Reducing power in the photolithotrophic bacteria (largely in the families Chlorobiaceae and Chromatiaceae) is provided by H_2S, S^0, thiosulfate, and H_2.

$$CO_2 + 2H_2S \xrightarrow{light} (CH_2O)_x + H_2O + 2S + NADH$$

$$CO_2 + 2S + 5H_2O \xrightarrow{light} 3(CH_2O)_x + 2H_2SO_4 + NADH$$

$$CO_2 + 2H_2 \xrightarrow{light} (CH_2O)_x + H_2O + NADH$$

The photoorganotrophic bacteria (largely in the family Rhodospirillaceae) strip hydrogen from organic compounds. For example, a species of *Rhodopseudomonas* turns isopropyl alcohol into acetone to provide reducing power for CO_2 fixation.

$$CO_2 + 2\ \overset{|}{\underset{|}{C}}-\overset{|}{\underset{OH}{C}}-\overset{|}{\underset{|}{C}}- \xrightarrow{light} (CH_2O)_x + \overset{|}{\underset{|}{C}}-\overset{}{\underset{O}{C}}-\overset{|}{\underset{|}{C}}-$$

The actual fixation of CO_2 into a carbohydrate takes place in a series of closely coupled metabolic reactions named after one of the workers, M. Calvin, who figured out the steps in the pathway. The initial step in fixing of CO_2 is catalyzed by ribulose diphosphate carboxylase. A six-carbon intermediate is formed which splits into two molecules of 3-phosphoglyceric acid (Fig. 5.4). In addition to the Calvin cycle, some photosynthetic bacteria also fix CO_2 in a reductive carboxylic acid cycle in which pyruvate, aspartate, α-ketoglutarate, and isocitrate are formed from the additions of CO_2 to precursor molecules (Fig. 5.5).

Chemotrophy. Nonphotosynthetic microbes obtain energy from the oxidation of inorganic compounds (chemolithotrophy) or of organic compounds (chemoorganotrophy).

CHEMOLITHOTROPHY. The oxidation of a number of inorganic compounds provides energy for the growth and reproduction of a diversity of chemolithotrophic bacteria. Each species of chemolitho-

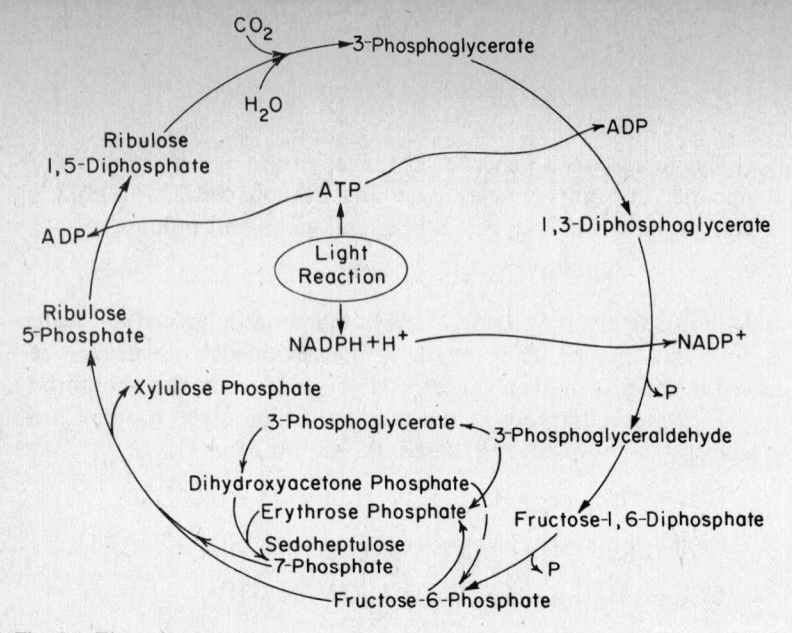

Fig. 5.4. The reductive pentose phosphate (Calvin-Benson) pathway.

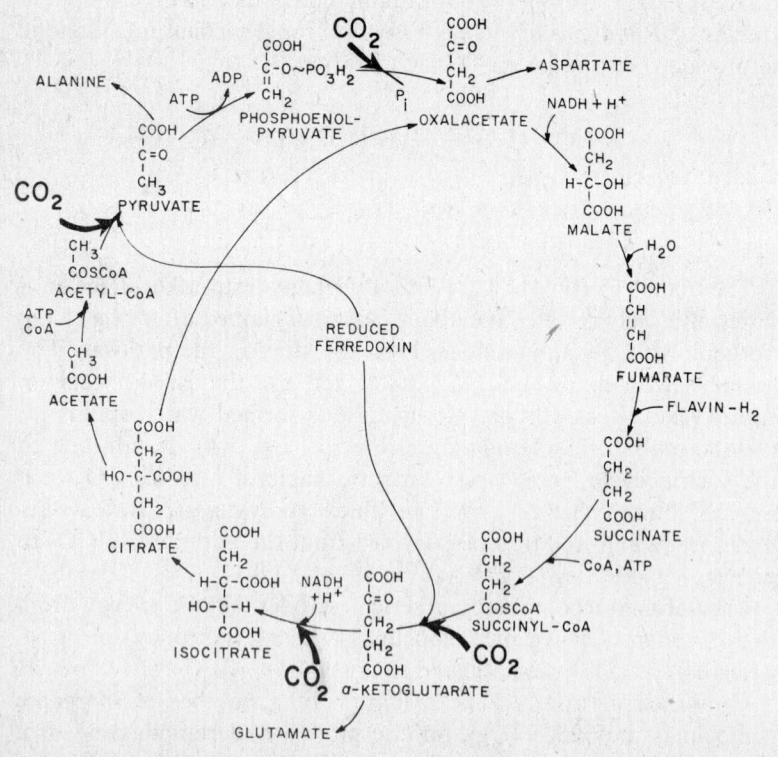

Fig. 5.5. The reductive carboxylic acid cycle used by some photosynthetic bacteria for the fixation of CO_2.

troph is quite restricted in substrate used as an energy source, and CO_2 is their only source of carbon for synthesis and growth. Examples are the sulfur- and iron-oxidizing, nitrifying, and hydrogen bacteria.

Hydrogenomonas species

$$H_2 + 1/2\ O_2 \rightarrow H_2O + 57\ \text{kcal/mole}$$

Thiobacillus thioxidans

$$S^0 + 3/2\ O_2 + H_2O \rightarrow H_2SO_4 + 118\ \text{kcal/mole}$$

Nitrosomonas species

$$NH_4 + 3/2\ O_2 \rightarrow NO_2^- + H_2O + 2H + 66\ \text{kcal/mole}$$

Thiobacillus ferrooxidans

$$4FeSO_4 + O_2 + 2H_2SO_4 \rightarrow Fe_2(SO_4)_3 + 2H_2O + 11\ \text{kcal/mole}$$

CHEMOORGANOTROPHY. Many nonphotosynthetic microbes, perhaps the majority of chemotrophs, derive energy for life functions from the oxidation of organic molecules. Although just about every organic molecule can be degraded by some microbial species for energy, there are metabolic pathways that serve as central cores for energy-liberating and biosynthetic reactions.

Fermentation. Fermentation is an ATP-generating process in which organic compounds serve as both electron donors and electron acceptors. Under these conditions (where oxidation of an organic compound takes place in the absence of an external electron acceptor), only a small amount of the potential energy of the substrate is liberated. Some of it is trapped in the production of one or two new ATP molecules.

$$C_6H_{12}O_6 \rightarrow 2CH_3CH_2OH + 2CO_2 + 2ATP + 32\ \text{kcal/mole}$$
glucose   ethanol   loss to system

There are four metabolic pathways that lead from glucose to pyruvic acid. The Emden-Meyerhof pathway requires the input of two high-energy bonds but yields 4ATP for a net gain of 2ATP for the entire pathway (Fig. 5.6). The three other pathways, the hexose monophosphate shunt, the Entner-Doudoroff pathway, and the phosphoketolase pathway, yield less energy—only one net ATP.

<u>Embden-Meyerhof Pathway.</u> The stepwise breakdown of glucose to pyruvate is called *glycolysis,* or the Embden-Meyerhof pathway (Fig. 5.6). It consists of an initial series of preparatory reactions

which do not involve oxidation but which lead to the formation of a key intermediate, glyceraldehyde-3-phosphate, and a second series of reactions in which oxidation-reductions occur, high-energy bonds are formed, and pyruvate is produced. From pyruvate, fermentation products characteristic of the particular microbe are released. In

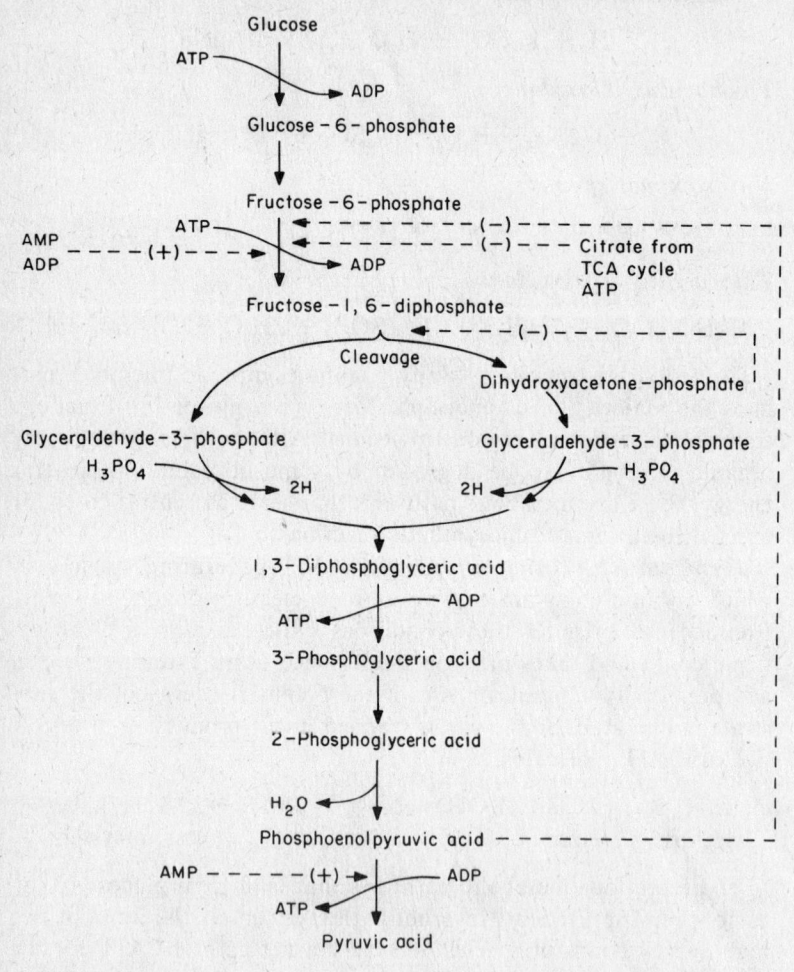

Overall reaction:
$$C_6H_{12}O_6 + 2\ ADP + 2\ H_3PO_4 \longrightarrow 2\ CH_3COCOOH + 2\ ATP + 2\ H_2O + 2\ (2H)$$
Glucose   Pyruvic acid

Fig. 5.6. The Emden-Meyerhof pathway (glycolysis) for the metabolism of glucose.

```
           Glucose
     ATP ─╮
          ╲
           ╰──▶ ADP
           │
           ▼
    Glucose-6-phosphate
           │
           ╰──▶ 2H
           ▼
    6-Phospho-glucono-lactone
     H₂O ─╮
           ▼
    6-Phospho-gluconic acid
           │
           ╰──▶ 2H
           ╰──▶ CO₂
           ▼
    Ribulose-5-phosphate
```

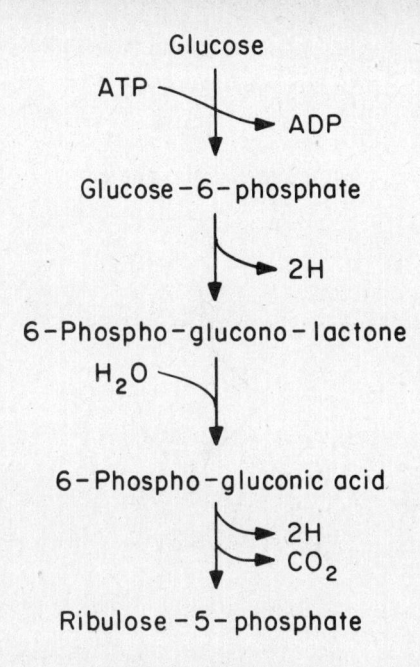

Fig. 5.7. The hexose monophosphate shunt which leads to the formation of ribose and deoxyribose needed for nucleic acid synthesis.

the initial steps one high-energy bond from ATP is used to transform glucose to glucose-6-phosphate which is isomerized to fructose-6-phosphate in the next metabolic step. A second ATP is used in the phosphorylation of fructose-6-phosphate to form fructose-1,6-diphosphate.

Hexose Monophosphate Shunt. Here glucose-6-phosphate is converted into 6-phosphogluconolactone and then oxidized to 6-phosphogluconic acid. In the next step, the 6-phosphogluconic acid is further oxidized and CO_2 is released, forming ribulose-5-phosphate (Fig. 5.7). This pathway has another vital function in metabolism. It produces ribose-5-phosphate, an important metabolite required for nucleic acid synthesis.

Entner-Douderoff Pathway. This pathway is similar to the hexose monophosphate shunt in its initial steps; it differs in the way in which 6-phosphogluconic acid is metabolized (Fig. 5.8). Instead of being oxidized as in the hexose monophosphate shunt, it is dehydrated to 2-keto-3-deoxy-6-phosphogluconic acid. The end products are pyruvic acid and glyceraldehyde-3-phosphate.

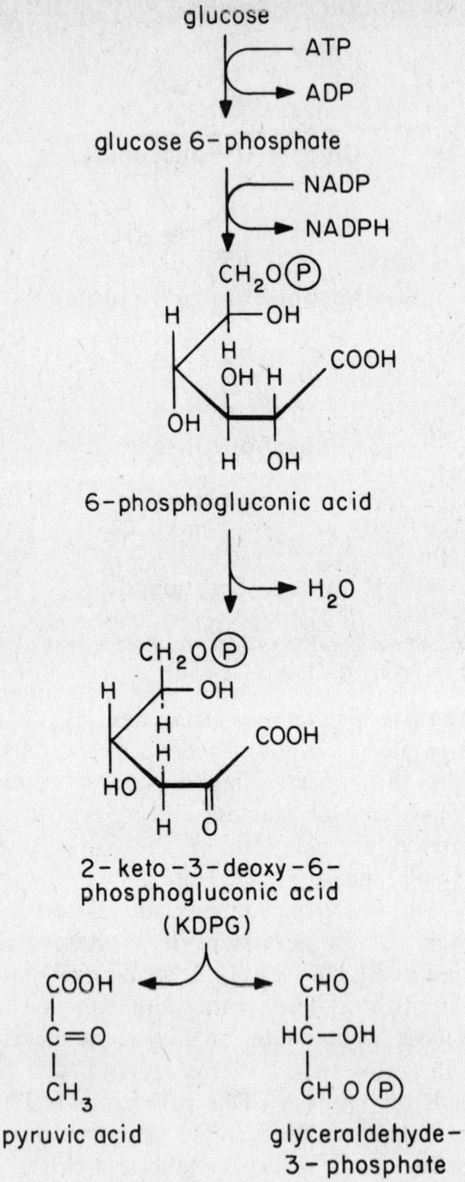

Fig. 5.8. The Entner-Douderoff pathway.

Basic Patterns of Energy Trapping

<u>Phosphoketolase Pathway.</u> In this pathway, either glucose or ribose can be oxidized to produce lactate, acetate, ethanol, and ATP. Glucose is decarboxylated to pentose; subsequently, xylulose-5-phosphate is formed. Phosphoketolase cleaves the xylulose-5-phosphate to glyceraldehyde-3-phosphate and the energized metabolite, acetyl-phosphate. From the latter two substances, lactate (through pyruvic acid) and acetate are formed, respectively.

The end products of bacterial fermentations and the pathways by which they are formed are characteristic for particular species, genera, and sometimes families. Pyruvic acid is a key metabolite from which many fermentation products are derived (Fig. 5.9). Among

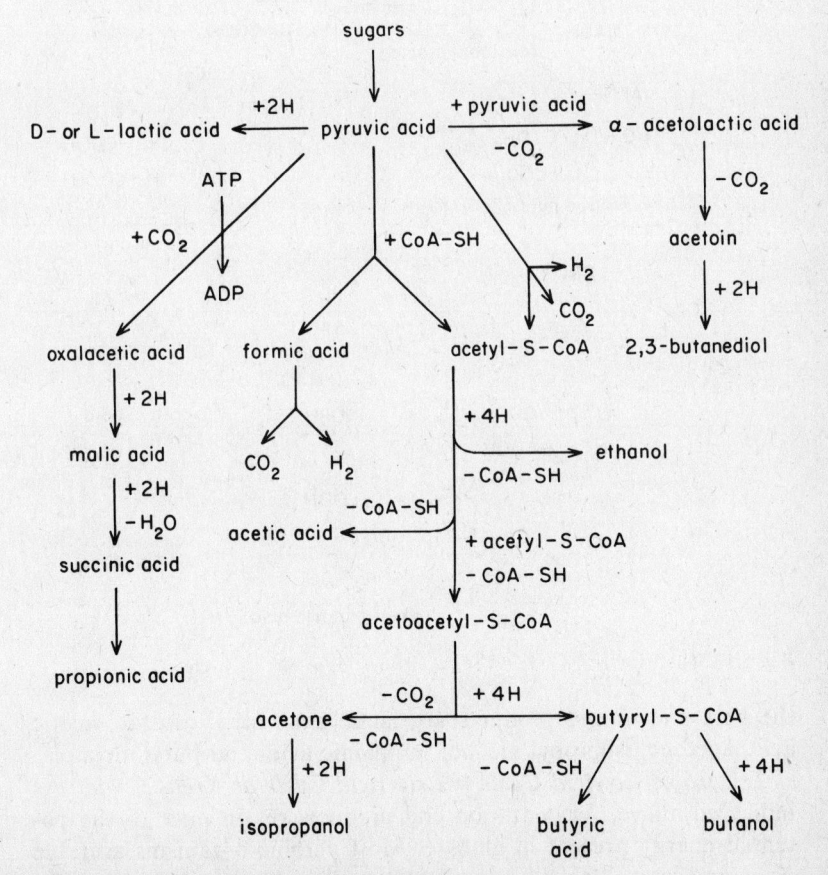

Fig. 5.9. End products of bacterial fermentations.

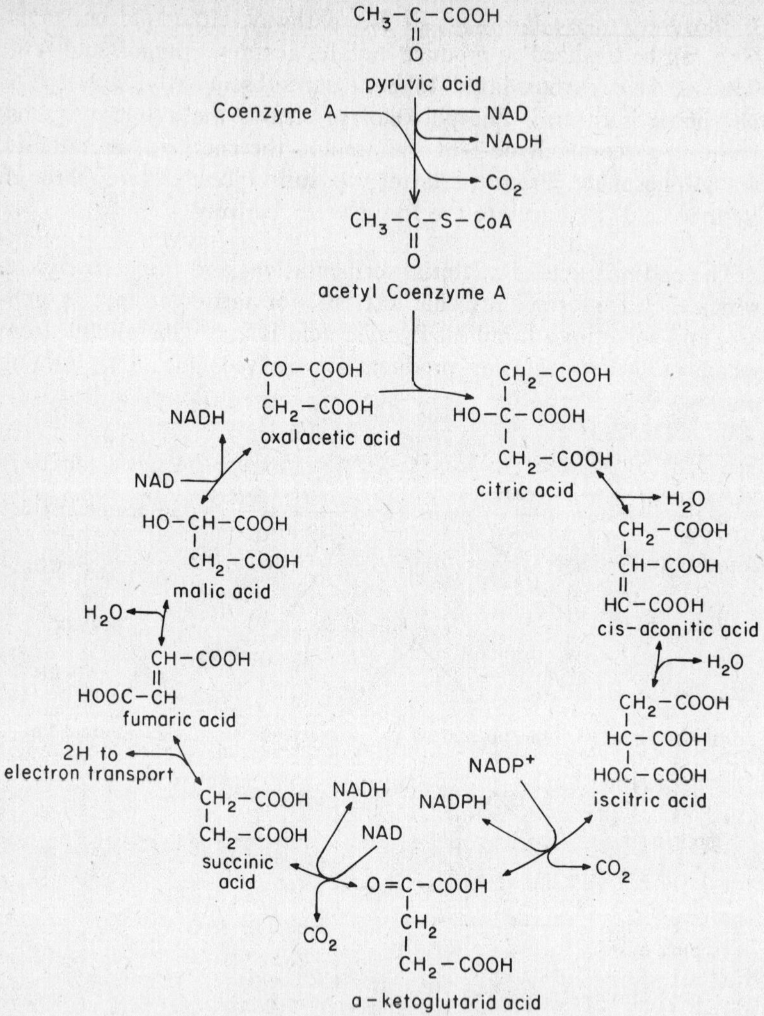

Fig. 5.10. The tricarboxylic acid (TCA, citric acid, or Krebs) cycle.

the best known end products are lactic acid, ethyl alcohol, acetic acid, acetone, isopropyl alcohol, propionic acid, and butyl alcohol.

Tricarboxylic Acid Cycle (Citric Acid Cycle or Krebs Cycle). As indicated above, fermentation end products retain most of the potential energy present in glucose. Most aerobic organisms are able to completely oxidize glucose to CO_2 and water through the metabolic pathway known as the tricarboxylic acid cycle. Molecules oxidized by the TCA cycle are funneled by one pathway or another to pyruvic acid. Pyruvic acid is decarboxylated and oxidized to form CO_2, the reduced coenzyme $NADH_2$, and an activated metabolite acetyl-coenzyme A (abbreviated acetyl-CoA). Acetyl CoA is an intermediate metabolite involved in many biosynthetic reactions. The acetyl group (two carbons) of acetyl-CoA enters the citric acid cycle (Fig. 5.10) by combining with the four-carbon compound oxalacetic acid to form citric acid, a six-carbon organic acid. Dehydration, decarboxylation, and oxidation reactions follow until oxalacetic acid is regenerated, completing the cycle. For each pyruvate molecule that is channeled through the TCA cycle, $3CO_2$, $4NADH_2$, $1FADH_2$, and one high-energy bond are generated. The reduced coenzymes are oxidized through the respiratory chain to yield a total of 15 ATP molecules synthesized for each turn of the cycle. For each glucose molecule completely oxidized through the Embden-Meyerhoff pathway and the TCA cycle, a total of 38 ATP molecules are formed.

Glyoxylate Cycle. A special modification of the TCA cycle is known as the glyoxylate cycle. It comes into play during the oxidation of acetic acid and higher fatty acids. In the glyoxylate cycle, oxalacetic acid required for acetic acid oxidation is generated from malic acid which is formed either from fumaric acid, the conventional substrate in the TCA cycle, and by the condensation of acetyl-CoA and glyoxylic acid (Fig. 5.11).

Respiration. Oxidation of an organic energy source with an external electron acceptor is called respiration. The most common external electron acceptor is gaseous oxygen, which is converted to water at the end of respiration. While only a fraction of the potential energy in a glucose molecule is captured by one of the fermentative pathways (a 1 or 2ATP net gain), a potential yield of 38 ATP molecules is possible if all the substrate molecules are oxidized to CO_2 and water. One of the keys to understanding the differences between respiration and fermentation is to follow the manner in which the reduced coenzyme $NADH_2$ is oxidized. The electrons from $NADH_2$, instead of being transferred to intermediates such as pyruvic acid or acetaldehyde in a terminal oxidation, are transferred to oxygen through the mediation of a respiratory chain. The respiratory chain is able to trap energy in the form of three high-energy bonds (ATP) for every reduced coenzyme molecule that is oxidized (Fig. 5.2).

Anaerobic Respiration. Some microorganisms have the ability to use molecules other than O_2 as electron acceptors in the oxidation of $NADH_2$ through respiratory chains. The process of oxidation with one of these alternate electron acceptors is known as anaerobic respiration. Denitrification is the best known and perhaps the most common process that uses an alternate electron acceptor. In the absence of O_2, denitrifying bacteria use NO_3^- which is reduced to NO_2^-, N_2O, or N_2 during respiration. Only 2ATP rather than 3 are

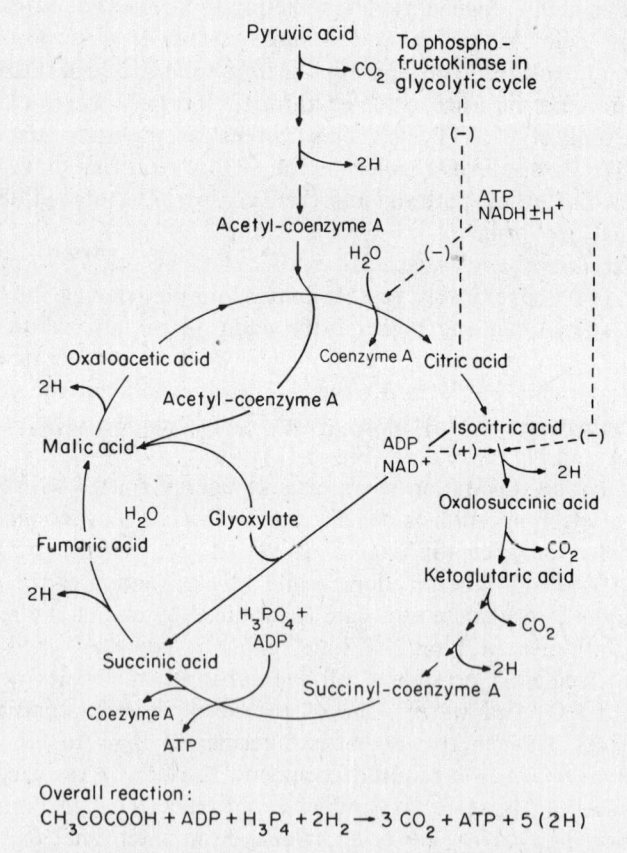

Overall reaction:
$$CH_3COCOOH + ADP + H_3P_4 + 2H_2 \rightarrow 3\ CO_2 + ATP + 5(2H)$$

Fig. 5.11. The glyoxylate modification (shunt) of the tricarboxylic cycle.

generated in the process of reducing NO_3^- to NO_2^- because the respiratory chain is shortened from that involving O_2. Some of the denitrifying bacteria can also oxidize $NADH_2$ by reducing ferric ion (Fe^{3+}) to ferrous (Fe^{2+}) under anaerobic conditions. Another electron acceptor used by some bacterial species is SO_4^{2-}. Sulfate-reducing bacteria are strict anaerobes and reduce SO_4^{2-} to H_2S. Another group of strict anaerobes using an alternative electron acceptor is the methanogenic bacteria. These bacteria, which live in sewage treatment plants, in certain types of soils, and in the rumen of herbivorous animals, use CO_2 as an electron acceptor, reducing it to methane (CH_4) during respiration.

Pasteur Effect. Many microorganisms have the enzymatic capability of carrying out fermentation under anaerobic conditions and respiration in the presence of O_2. For example, if oxygen is bubbled into a yeast suspension fermenting glucose, glucose consumption slows down and alcohol production stops. This phenomenon was first observed by Pasteur. We now know the considerable energetic benefits of the Pasteur effect to organisms that have such metabolic switching capabilities. Less glucose is needed for aerobic respiration than for fermentation, since more ATP (38 versus 2ATP) is generated per mole of substrate. Recent metabolic studies indicate that the levels of cellular $NADH_2$ and ATP/ADP exert controls on the pathway followed by the cell.

BIOSYNTHESIS AND METABOLIC REGULATION

Biosynthesis (anabolism) is the process by which cell constituents are built up from simpler starting materials. The energy for biosynthesis, when required, comes from ATP. Most of the organic constituents of the cell belong to four major classes of macromolecules: polysaccharides, proteins, nucleic acids, and complex lipids. Approximately 150 different molecules go into the construction of a new cell. The precursors of the macromolecules are synthesized in anabolic pathways that are similar in all organisms. Required for this synthesis are energy from stored ATP and the basic elements of biological compounds: carbon, hydrogen, oxygen, nitrogen, sulfur, and phosphorus. These elements are obtained from inorganic or organic sources. Precursor compounds that cannot be synthesized by a particular organism must be present in the immediate environment. Such substances, known as growth factors, are vitamins, amino acids, purines, and pyrimidines.

The main metabolic pathways are *amphibolic*—that is, they function both in anabolism and catabolism. In eucaryotes, anabolic and catabolic reactions are most often separated from each other in different parts of the cell. Many catabolic reactions are localized in mitochondria and microbodies while anabolic reactions are associated with cytoplasmic organelles.

Pathways like the tricarboxylic acid cycle cannot continue to function anabolically without help; soon, key compounds would be used up, and there could not be a repeating cycle. The glyoxylate shunt (Fig. 5.11) is a mechanism to regenerate oxalacetate and keep the TCA cycle going. *Anaplerotic* is the general term for reactions that serve to replenish compounds needed to keep metabolic pathways functioning.

Methods of Studying Biosynthetic Pathways

AUXOTROPHIC MUTANTS. The concept of systematic isolation of *auxotrophic* mutants (organisms that require growth factors which the parental strains can synthesize) developed by Nobel laureates Beadle and Tatum was a major advance in the study of biosynthetic pathways. Nutritional mutants usually have lost the ability to synthesize an enzyme catalyzing a step in a metabolic pathway. Information on the sequence of reactions in a pathway can be obtained by testing the ability of mutant strains to grow on suspected intermediates of the pathway being investigated. The pathway may also be studied by finding any metabolic intermediates that may have accumulated in mutants because of a missing enzyme.

RADIONUCLIDE TRACER METHODS. Radioisotopes are commonly used to study metabolic pathways. In the technique known as *pulse labeling,* a culture is briefly exposed to a radioactively labeled biosynthetic precursor. During the brief exposure, the precursor enters the cell and the metabolic pathways in which it participates. At various times after the introduction of the radioactive tracer-labeled metabolite, cell samples are chemically fractionated and analyzed, revealing the sequence of chemical transformation in the pathways leading from the precursor molecules.

Assimilation of Elements. Carbon, hydrogen, oxygen, nitrogen, sulfur, and phosphorus are the most abundant elements in cell structure. Oxygen and hydrogen are easily obtained from the water that bathes every cell. As discussed earlier in this chapter, phototrophs generally fix CO_2 in the Calvin cycle by a reaction catalyzed by the enzyme ribulose diphosphate carboxylase (Fig. 5.4). Chemotrophs can also fix CO_2. In one pathway CO_2 is fixed into the carboxyl group of oxalacetate, an intermediate of the tricarboxylic acid cycle (Fig. 5.5). Methanogenic bacteria, which are anaerobes, use CO_2 as a terminal electron acceptor, converting it to methane (CH_4) and water.

ASSIMILATION OF INORGANIC NITROGEN. Although gaseous nitrogen is abundant and is soluble in water, only a very few specialized microorganisms can assimilate it. Ammonia and nitrate, on the other hand, are readily assimilated by most species. However, not all the species that grow readily with ammonia as a nitrogen source can use nitrate; such species do not possess the necessary enzymes or co-factors.

Nitrate and Nitrite Assimilation. If nitrate is used as a source of nitrogen, it is converted first to ammonia, a reductive metabolic reaction in which the valence of the nitrogen atom is changed from +5 to −3. Assimilatory nitrate reduction is a two-step process, involving two enzyme complexes, nitrate reductase and nitrite reductase, and pyridine nucleotides (NADH or NADPH) which supply electrons. The reductase enzyme complexes have two high molecular weight proteins, one of which functions with FAD or ferredoxin to accept the electrons from reduced pyridine nucleotides and the other of which contains molybdenum (Mo) which effects the reduction of nitrate to nitrite, or nitrite to ammonium.

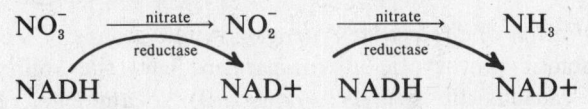

$$NO_3^- \xrightarrow[\text{reductase}]{\text{nitrate}} NO_2^- \xrightarrow[\text{reductase}]{\text{nitrate}} NH_3$$

NADH NAD+ NADH NAD+

Certain microorganisms are able to use nitrate as a terminal electron acceptor for anaerobic respiration.

Assimilation of Ammonia. The nitrogen atom in ammonia is at the same oxidation level as it is in the organic constituents of cells. Ammonia is assimilated in a reaction involving an exchange with the keto group of α-ketoglutaric acid, forming the amino group of glutamic acid, or an exchange with the terminal hydroxy group of either aspartic or glutamic acids, forming the amido group of asparagine or glutamine, respectively.

(1)

$$NH_3 + HOOC(CH_2)_2-\overset{O}{\underset{}{C}}-COOH \xrightarrow[NADH + H \quad NAD^+]{\text{glutamate dehydrogenase}}$$

$$HOOC(CH_2)_2-\underset{}{\overset{NH_2}{C}}-COOH + H_2O$$
$$\text{glutamic acid}$$

(2) $NH_3 + HOOC-CH_2-CHNH_2-COOH \xrightarrow[ATP \quad AMP + \text{\textcircled{P}} - \text{\textcircled{P}}]{\text{asparagine synthetase}}$
       aspartic acid

$$\overset{O}{\underset{NH_2}{C}}-CH_2CHNH_2-COOH$$
$$\text{asparagine}$$

Glutamic acid and glutamine play important roles in amino acid synthesis by transferring their amino or amido groups to the precursors of many amino acids and nucleic acids.

(3) Glutamic acid + pyruvic acid → α-ketoglutaric acid + alanine

(4) Uridine triphosphate + glutamine ──────→
         cytidine triphosphate + glutamic acid
    ATP ────→ ADP + $\text{\textcircled{P}}$

Assimilation of Molecular Nitrogen. A few groups of free-living and symbiotic procaryotic microorganisms have the ability to fix (reduce) gaseous nitrogen, N_2 (valence 0), to ammonia. The enzyme system responsible, nitrogenase, consists of two separate iron-containing components. Component I, which binds the N_2 also contains molybdenum. Component II is activated by ATP. It accepts electrons from ferredoxin and transfers them to the N_2 bound to component I. Even in aerobes the nitrogenase components are sensitive to O_2, suggesting that the process is carried out in an O_2-protected microenvironment. In *Clostridium pasteurianum,* one of the best studied free-living, anaerobic, soil-dwelling nitrogen fixers, ferredoxin is reduced by the splitting of pyruvate; ATP required for nitrogen fixation is synthesized at the same time.

(1) Phosphoclastic reaction:

$$CH_3COCOOH + H_3PO_4 \xrightarrow{\text{pyruvate dehydrogenase}}$$
pyruvic acid

$$CH_3(CO)H_2PO_4 + CO_2\uparrow + 2H^+$$
acetyl phosphate → reduced ferredoxin → electrons

CH_3COOH + ATP
acetic acid

(2) Nitrogen activation and reduction:

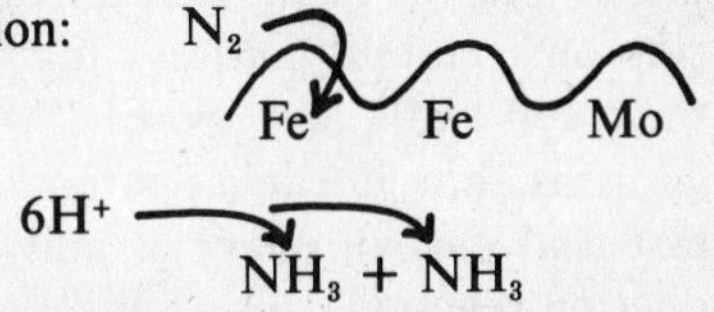

N_2 → Fe — Fe — Mo

$6H^+ \longrightarrow NH_3 + NH_3$

ASSIMILATION OF SULFUR. Sulfur is needed for the synthesis of two amino acids (cysteine and methionine) and several vitamins (thiamine, biotin, and α-lipoic acid). Most microorganisms obtain sulfur by reducing sulfate (valence +6) to sulfide (valence −2). Several steps are involved: reactions with ATP to form phosphoadenosine phosphosulfate (PAPS); reduction of the sulfate radical in PAPS to sulfite; reduction of the sulfite radical to H_2S. The reductions use NADPH as an electron donor. Organic sulfur compounds are formed by reaction of precursor molecules like serine with H_2S.

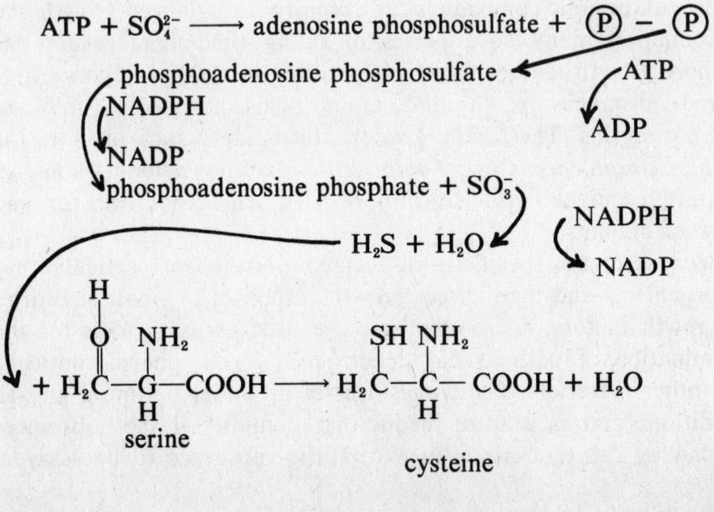

$$ATP + SO_4^{2-} \longrightarrow \text{adenosine phosphosulfate} + \text{P} - \text{P}$$

phosphoadenosine phosphosulfate ← ATP
NADPH ↓ ADP
NADP
phosphoadenosine phosphate + SO_3^-
$H_2S + H_2O$ ← NADPH → NADP

$$+ H_2\overset{H}{\underset{H}{\overset{|}{C}}}-\overset{NH_2}{\underset{|}{C}}-COOH \longrightarrow H_2\overset{SH}{\underset{|}{C}}-\overset{NH_2}{\underset{H}{\overset{|}{C}}}-COOH + H_2O$$
serine → cysteine

ASSIMILATION OF PHOSPHORUS. Phosphorus is needed for the synthesis of phospholipids and nucleic acids. Most microorganisms readily assimilate inorganic and organic phosphates.

MINERAL NUTRITION. A small number of minerals are required for growth and reproduction of all living organisms.

Macronutrients. Minerals required in relatively large amounts are potassium, magnesium, calcium, sodium, and iron. *Potassium* activates a variety of enzymes. *Magnesium* also activates enzymes, and it is a key atom in the chlorophyll molecule. Magnesium complexes with nucleotides, particulary with ATP, and is important for ribosome stability. Many microorganisms concentrate *calcium* in their cell walls and spores. *Sodium* is required by many microorganisms, but it can apparently be replaced in others. *Iron* is an essential part of enzymes and coenzymes involved in oxidation-reduction reactions, where it serves as an electron carrier. Radiolaria, diatoms, and microorganisms with *silicon* in their cell walls require large amounts of this element.

Micronutrients (Trace Elements). Micronutrients, or *trace elements,* are minerals required in very small amounts. The best-known trace elements are zinc, copper, cobalt, manganese, and molybdenum. These minerals also function as parts of enzymes and coenzymes.

Organic Growth Factors. Specific organic compounds required in small amounts by cells that cannot synthesize them are called growth factors. Although there are many microorganisms that synthesize all organic compounds they require for balanced growth and reproduction, many more species, including some algae, require one or more growth factors. Frequently falling into this category of required substances are vitamins, amino acids, purines, pyrimidines, and porphyrins. The family Lactobacteriaceae, which includes the genera *Streptococcus* and *Lactobacillus,* is known for its complex vitamin requirements—vitamin requirements that even exceed those of humans.

Because of the specificity of the requirements of particular microorganisms and their direct growth responses to small quantities of growth factors, microorganisms are used as assay tools for the examination of pathological deficiencies, foods, pharmaceuticals, and other materials. The assays are set up under optimum growth conditions and in culture media that contain all the substances needed by the test organism except the substance to be assayed.

Under these conditions growth is proportional to the amount of the limiting growth factor present in the material assayed.

Synthesis of Amino Acids. The building blocks of proteins are the 20 or more common amino acids. Except for histidine, the multiply branched metabolic pathways leading to the synthesis of all amino acids begin from a relatively small number of "central," or "key," intermediary metabolites in the Embden-Meyerhof (glycolytic) and tricarboxylic acid pathways. Depending upon their origin from particular central intermediary metabolites, amino acids are grouped into five families.

GLUTAMATE FAMILY. Glutamate, the initial compound of the glutamate family, was mentioned in context with ammonia assimilation. Other members of the glutamate family, which has its origin at α-ketoglutarate of the TCA cycle, are glutamine, arginine, proline, and sometimes lysine.

ASPARTATE FAMILY. The synthesis of the aspartate family (Fig. 5.12), which includes aspartic acid, asparagine, methionine, threonine, isoleucine, and lysine, begins with the transamination of another key TCA cycle intermediate, oxalacetic acid.

AROMATIC AMINO ACID FAMILY. Tryptophan, phenylalanine, and tyrosine are synthesized from the metabolite phosphoenolpyruvate, an energetic intermediary compound in the Embden-Meyerhof pathway, and erythrose-4-phosphate.

SERINE FAMILY. This family, which includes serine, glycine, and cysteine, is synthesized in a metabolic pathway which begins at 3-phosphoglycerate, another Embden-Meyerhof intermediary metabolite.

PYRUVATE FAMILY. The three amino acids belonging to the pyruvate family, alanine, valine, and leucine, are synthesized from pyruvate, a key metabolite intermediate produced at the end of the Embden-Meyerhof pathway.

HISTIDINE. A complex metabolic pathway that branches from the hexose monophosphate pathway is responsible for the synthesis of the amino acid histidine and purines.

Synthesis of Purines and Pyrimidines. Purines and pyrimidines are components of nucleic acids, some vitamins, and coenzymes. The purine and pyrimidine rings by themselves are called *free bases*. The base with a sugar (a pentose) attached is called a *nucleoside*. Depending upon the sugar (ribose or deoxyribose) the nucleoside is a *riboside* or a *deoxyriboside*. When a phosphate is attached to the sugar, the compound is known as a *nucleotide*. Nucleotides are symbolized by the letters A, G, U, C, or T to indicate which purine or pyrimidine base they contain. The ability to synthesize purines and pyrimidines is widespread among animals, plants, and microorganisms. Some microorganisms (e.g., the lactic acid bacteria and the trichomonads) lack this ability.

PYRIMIDINES. The six-membered ring of the pyrimidine skeleton, *orotic acid,* is synthesized by a condensation between the amino acid aspartic acid and carbamyl phosphate.

PURINES. The nine-membered ring is built up on a two ribose-phosphate unit by successive additions of amino groups and single carbon-containing groups transferred by the coenzyme folic acid. The first purine ring synthesized is that of *isosinic acid,* an intermediate in the formation of both *guanylic* and *adenylic acids*.

MACROMOLECULAR SYNTHESIS

Polysaccharides, proteins, and nucleic acids are *polymers,* that is, compounds or repeating subunits, or *monomers,* which are joined by bonds that are characteristic of each class of macromolecule. Their synthesis requires chemical activation of the monomer and the removal of a molecule of water.

Protein Synthesis. Bacterial cells have the genetic capability of synthesizing several thousand different kinds of proteins, each built up from an average of about 200 amino acid residues (monomers) linked together in definite sequences. The amino acid sequence of each protein is specified by a portion of the DNA molecule which, by definition, is called a *gene*. A sequence of three nucleotide bases, called a *codon,* is the genetic code for each amino acid. A number of amino acids are coded by more than one codon.

Protein synthesis involves two processes, transcription and translation.

TRANSCRIPTION. Transcription is the "passing" of information from the DNA nucleotide bases to *messenger RNA (mRNA)*. An enzyme called *RNA polymerase* synthesizes a sequence of ribonucleotide bases into an RNA molecule complementary to the DNA sequence being transcribed; DNA serves as a template.

TRANSLATION. Translation is the process by which the message carried by mRNA is translated into the amino acid sequence of a protein. It involves another set of RNA molecules called *transfer*

RNA (tRNA). In a reaction using energy from ATP and catalyzed by an activating enzyme, each amino acid attaches to a specific tRNA. The sequence of three nucleotides on the tRNA, the *anticodon,* is complementary to mRNA coding for the particular amino acid bound to the tRNA.

Actual protein synthesis takes place on a *ribosome,* a cytoplasmic body composed of *ribosomal RNA (rRNA)* and protein. The active bacterial ribosome has a sedimentation velocity of 70S which can be reversibly dissociated into subunits of 30S and 50S.

The actual process of protein synthesis—translation—takes place in discrete steps: initiation, elongation, termination-release, and polypeptide folding.

Initiation. The *initiation complex* is formed by the union of a 30S ribosomal subunit and an animoacyl-tRNA enzyme on a specific site of a mRNA molecule. With the aid of three accessory proteins (*initiation factors*) and expenditure of energy from the high-energy bond of guanosine triphosphate (GTP), a 50S ribosomal subunit attaches to the complex, forming the functional 70S ribosome.

Elongation. Elongation, or the addition of each amino acid during the polymerization, is similar to the initiation process. It involves three events: recognition, peptidyl transfer, and translation.

<u>Recognition.</u> A molecule of aminoacyl-tRNA attaches to a recognition site (A site) on the 70S ribosome, which exposes a codon on the mRNA. The amino acid coded for at the A site is the one that will be added to the peptide chain growing nearby at the P (peptide) site. Two accessory proteins (elongation factors) and energy from GTP are required in the recognition and attachment process.

<u>Peptidyl Transfer.</u> The peptide at the P site is then transferred and joined by a peptide bond to the amino acid attached to the tRNA molecule at the A site, thereby lengthening the peptide chain by one residue.

<u>Translocation.</u> Following the peptidyl transfer, the peptide-bearing tRNA is moved back (translocated) to the P site along with the mRNA, and the next codon is exposed at the A site. Translocation requires an accessory protein (elongation factor) and energy from the hydrolysis of GTP.

Termination-release. The process of recognition, peptidyl transfer, and translocation continues until a nonsense codon (UAG, UAA, or UGA) is reached which signals completion and causes the release of the peptide from the ribosome. Another protein, the release factor, mediates this process. After the release, the 70S ribosome dissociates into its 30S and 50S subunits.

Polypeptide folding. Before the release, the polypeptide chain forms an α-helix (secondary conformation) and begins to fold on itself (tertiary conformation). The folding is determined by the specific amino acid sequence. In most proteins, much of the tertiary structure is maintained by disulfide bridges and hydrophobic bonds.

DNA Synthesis. The double helical structure of DNA molecules with their linear sequences of nucleotide pairs is well known to all biologists. Each nucleotide has a 5' phosphate end and a 3' hydroxyl end. A polarized chain with a 5' phosphate end and a 3' hydroxyl end is formed in the process of linking together a series of nucleotides by phosphodiester bonds. The two strands of the double helix run in opposite directions, that is, they are antiparallel.

INITIATION OF SYNTHESIS. With the aid of three enzymes, *DNA polymerases,* DNA synthesis proceeds in the 5' to 3' direction by the sequential formation of phosphodiester bonds between the α-phosphates of 5' deoxynucleoside triphosphates and the terminal 3' hydroxyl group of the DNA molecule (primer) to which the nucleotides are being attached. One pyrophosphate molecule is released for each deoxynucleotide added. The strand of the molecule opposite the primer is the template that determines the sequence of addition of deoxynucleotides.

STRAND SEPARATION. The separation of the double helix into single strands of DNA to allow for polymerization is somewhat more complicated. With the aid of a DNA-dependent RNA polymerase, a short complementary piece of RNA is synthesized on each of the separated DNA strands. The RNA acts as an initial primer.

POLYMERIZATION. Starting from the initiation point on the bacterial chromosome, polymerization takes place simultaneously around the bacterial chromosome in both directions, creating two forks of replication. After about 100 nucleotides have been polymerized onto the RNA primer by DNA polymerizase (Pol III), a second polymerase (Pol I) hydrolyzes the RNA fragment and polymerizes more DNA in its place. The relatively short pieces of DNA polymerized are linked together, by another enzyme, *DNA ligase,* to form continuous complementary copies of the original DNA strands.

RNA Synthesis. The synthesis of all three classes of RNA (mRNA, tRNA, and rRNA) is similar.

INITIATION OF SYNTHESIS. Single strands of DNA complementary to RNA are polymerized from the four ribonucleoside triphosphates (ATP, GTP, CTP, and UTP) with the aid of the complex enzyme, DNA-dependent RNA polymerase. In the formation of RNA, uracil rather than thymine pairs with adenine. Short specific sequences of DNA base pairs called *promoters* determine the points on one of the strands of the bacterial genophore that will be transcribed. The strand that is copied is not the same in all regions of the genophore.

POLYMERIZATION. As in DNA synthesis, polymerization of RNA occurs by forming phosphodiester bonds between the $3'$ hydroxyl group on the growing end of the RNA strand and the α-phosphate group of a free $5'$ nucleoside triphosphate.

TERMINATION OF TRANSCRIPTION. Polymerization continues until one or more genes have been transcribed. The end of transcription is signaled by a sequence of base pairs in the DNA that causes release of the RNA polymerase. The RNA formed is single stranded, but in the case of tRNA it always folds back on itself with the formation of hydrogen bonds between complementary bases to produce double-stranded regions.

METABOLIC REGULATION

In the course of metabolism, living cells take up molecules from their environment, capture and release energy, and direct the orderly flow of precursor compounds along multiple biosynthetic pathways to provide adequate supplies of each critical end metabolite. Several mechanisms have evolved which conserve energy and allow cells to coordinate their metabolic activities in response to available energy sources and needed precursor molecules.

Induction of Enzyme Synthesis. Although each microbial species has the genetic capacity to synthesize a wide range of metabolic enzymes, all enzymes are not present in cells at the same time. Thus cells do not waste materials and energy synthesizing enzymes that are only useful in particular environmental situations. (Certain enzymes, *constitutive enzymes,* necessary for metabolism under all growth conditions are always present in cells.)

GENERAL INDUCIBLE ENZYMES. *Inducible enzymes* are those enzymes produced only in response to the presence of particular metabolites in the environment. The best-known example of this kind of regulation of enzyme synthesis is that of the enzymes associated with the utilization of lactose, first studied by Nobel Prize Winner J. Monod and his colleagues. When *E. coli* is grown in media containing glucose, the cells contain only barely detectable levels of the enzymes necessary for the initial steps in lactose metabolism. When lactose is introduced into the medium, it induces the synthesis of several enzymes: *galactoside permease,* which mediates entry of lactose into cells, and *β-galactosidase,* which cleaves lactose into glucose and galactose. Galactose, in turn, induces the formation of the enzymes responsible for the metabolism of galactose (galactokinase, galactose transferase, and galactose epimerase).

COORDINATE SYNTHESIS. The synthesis of two or more metabolically linked enzymes is known as coordinate synthesis. For example, compounds that induce the formation of β-galactosidase also induce the synthesis of galactoside permease. Coordinacy is usually the result of the close association on the genophore of the structural genes encoding the metabolically linked enzymes.

MECHANISM OF INDUCTION. The induction of enzyme synthesis is mediated by particular allosteric proteins called *repressors* synthesized by specific regulatory genes. Repressors bind to the bacterial genophore at a site near inducible structural genes and prevent transcription. Inducing substances (inductors) bind to the repressor, alter its structure, and hinder its ability to block transcription (derepress). Following this, the transcription of the structural gene proceeds without hindrance.

End Product Inhibition (Feedback Inhibition; Retroinhibition). Many enzymes, particularly those that catalyze the first reaction in a biosynthetic pathway, are also allosteric proteins, that is, proteins whose properties are modified if small molecules called *effectors* are bound to them. As unused metabolites accumulate at the end of the pathway, they act as effectors, *slowing down the activity* of the enzyme(s) that leads to their formation.

End Product Repression. Accumulation of metabolites can also result in the *slowing down or complete arrest of enzyme synthesis* early in the metabolic pathway, leading to the formation of the surplus metabolite. End-product repression and end-product inhibition are complementary mechanisms (Fig. 5.12). The former regulates enzyme concentrations and the latter the activity of enzymes

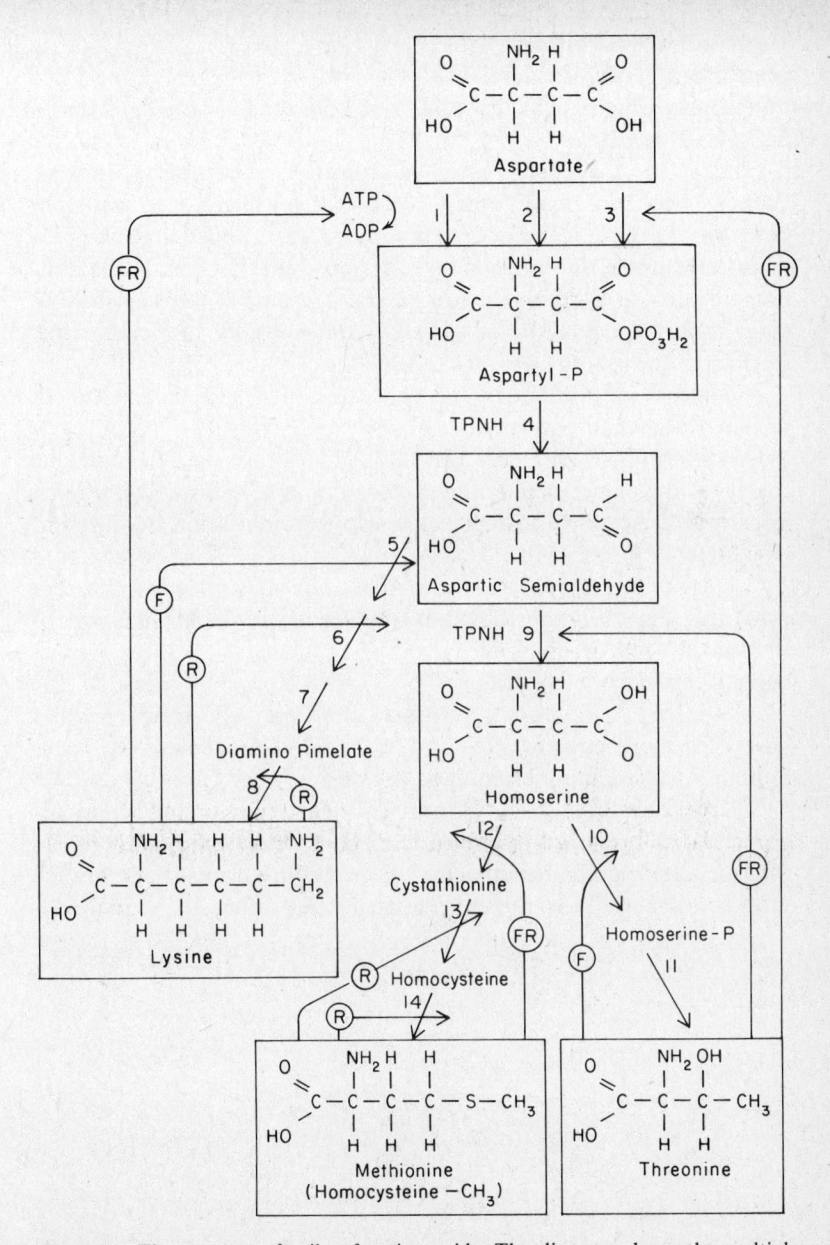

Fig. 5.12. The aspartate family of amino acids. The diagram shows the multiple branched pathways for the synthesis of individual amino acids and mechanisms (R = repression; F = feedback inhibition) which direct the molecules along the pathways to provide adequate supplies of each metabolite.

already present. This complementary action enables cells to provide adequate supplies of each critical end metabolite while at the same time saving precursor compounds and energy that can be diverted for needed syntheses.

Control in Branched Pathways. Many biosynthetic pathways, such as the aspartate family synthetic pathway, are multiply branched (Fig. 5.12). Such pathways present more control problems than simple unbranched ones. If accumulation of a single amino acid such as threonine could repress the entire pathway leading from aspartate, it could also affect the supplies of homoserine, methione, and lysine.

A number of inhibitory mechanisms are found in fine-tuned, multiply branched pathways:

ISOFUNCTIONAL ENZYMES (ISOENZYMES). More than one enzyme catalyzes the same reaction at the beginning of the pathway. Each isoenzyme is sensitive to repression and retroinhibition by different effectors.

CONCERTED FEEDBACK INHIBITION. This occurs when the enzyme has two different allosteric sites, each of which binds one of the specific end products of the pathway. Unless both sites are occupied, the enzyme is active.

CUMULATIVE FEEDBACK INHIBITION. Some allosteric enzymes have even more effector sites and all the end products must accumulate for complete inhibition to take place.

SEQUENTIAL FEEDBACK INHIBITION. This type of inhibition occurs in some branched metabolic pathways where the buildup of an intermediate metabolite immediately preceding a metabolic branch inhibits the activity of the enzyme at the beginning of the pathway.

Chapter 6

VIRUSES

It is often argued that because viruses lack some of the basic characteristics of life (i.e., movement, irritability, and growth), they cannot by definition be considered alive. Little is to be gained from such an argument. Viruses possess such extremely potent complements of genes that they can be replicated with great fidelity by living systems. If we consider viruses to be extremely complex macromolecules which have the inherent information to direct their own replication, we are almost obliged to conclude that a virus comes to life at the time it infects a cell.

GENERAL PROPERTIES

Quite a variety of unrelated genetic elements are collectively categorized in the group we recognize as viruses. They share the following properties:

1. They usually contain a single molecule of either DNA *or* RNA. (Some large RNA viruses contain more than one molecule.)

2. They can alternate between intracellular and extracellular states. The extracellular state, called a *virion,* consists of one or more molecules of nucleic acid surrounded by a protein coat (capsid).

3. They have varied by reduced genetic capacities (10 to several hundred genes), fewer genes than procaryotes, and lack the genetic coding for most metabolic pathways, particularly energy-liberating ones. They also lack ribosomes. Therefore, viruses must rely on host cells for all the energy and metabolic machinery needed to synthesize the components for virus replication.

4. They are very small and pass through filters designed to retain bacteria ($\sim 0.4 \mu m$).

FORM AND CHEMICAL COMPOSITION OF A VIRION

A virion consists of a nucleic acid core and a protein coat. Some animal virions are enclosed within an additional membranous envelope called a *mantle* or *peplos*. The mantle is derived from the cell membrane of the host cell; often the normal membrane has been modified through the action of viral genes.

Nucleoid or Core. In some viruses the nucleic acid molecule is wound into a ball or some other form called a nucleoid or core. Nucleic acids constitute from 1 percent (influenza virus) to around 50 percent (some bacterial viruses) of the total weight of viruses. A virus contains one of four possible types of nucleic acid: single-stranded DNA, double-stranded DNA, single-stranded RNA, and double-stranded RNA. Most viruses contain either double-stranded DNA or single-stranded RNA. Single-stranded DNA is found in some bacterial and animal viruses (e.g., coliphages FD and ϕX174). Double-stranded RNA is found in reoviruses and some plant viruses. The DNA of some viruses (T-even bacterial viruses) has an unusual base, 5-hydroxymethylcystosine, in place of cytosine.

Capsid. The proteinaceous coat surrounding the core of viruses is made up of many small subunits called *capsomeres*. The capsid and its nucleic acid core is sometimes referred to as the *nucleocapsid*. Though antigenic, the capsid is believed to be physiologically inert. Viruses fall into two basic groups: helical or polyhedral, based on the way their capsomeres are organized. A few viruses, such as some bacterial viruses (bacteriophages), have both kinds of symmetry—a polyhedral "head" joined to a helically constructed "tail."

HELICAL SYMMETRY. This form resembles a long cylinder with a hollow center. The capsomeres are bean-shaped and attached to a long, helically wound molecule of RNA. The best-known example of a virus with helical symmetry is the *tobacco mosaic virus*.

ICOSAHEDRAL SYMMETRY. Here the capsomeres form an icosahedron, a polyhedron with 12 corners, 20 identical triangular faces, and 30 edges. The capsids of polyhedral virions are composed of two types of capsomeres: hexamers, which fill the triangular faces, and pentamers, which are situated at the corners. The difference between the two types of capsomeres is the number of subunits in each; pentamers have five polypeptide chains and hexamers have six. The largest known icosahedral virion, from an insect, contains 1472 capsomeres. Adenovirus (type 2) and human wart viruses are well-known examples of viruses with cubical or polyhedral symmetry.

BINAL (COMPLEX) SYMMETRY. Helical and polyhedral structures are combined. Bacteriophages T2, 4, and 6 from *E. coli* are the best examples of binal viruses. In these forms the bipyramidal hexagonal prismatic head that contains the DNA is similar in construction to the capsid of polyhedral viruses. The tail is a hollow central tube surrounded by a sheath of contractile proteins. At the end of the tube is a hexagonal *base plate* with six long tail fibers attached at every corner. Other complex bacteriophages vary in the details of their construction.

Viroids. These are small RNA molecules (molecular weight 75,000 to 100,000) without capsids which are responsible for a small number of plant diseases (e.g., potato spindle tuber disease) and animal diseases (e.g., scrapic disease of sheep).

CLASSIFICATION

The LHT system, an acronym for Drs. Lwoff, Horne, and Tournier, the scientists who developed it, divides viruses into smaller categories on the basis of (1) type of nucleic acid, (2) symmetry, (3) presence or absence of an envelope around the nucleocapsid, (4) the detailed structure of the capsid and capsomeres, and (5) host-virus interactions. Unlike most classification schemes, the LHT system does not imply any evolutionary or phylogenetic relationships. At the present time, these relationships are obscure. Table 6.1 summarizes viral classification based on the LHT system.

BACTERIAL VIRUSES (BACTERIOPHAGES)

Although viruses have been found in cells of almost every group of living organisms, the viruses of bacteria (bacteriophages) have always been the special focus of viral research. Both DNA and RNA phages are known. As a general rule, the nucleic acid of DNA phages is double stranded and that of RNA phages is single stranded. A few cases of phages with single-stranded DNA are known. The capsid of almost all phages is polyhedral; many have "tails." Some helical, or filamentous, phages (example, fd) contain-

Table 6.1 Classification of Viruses (LHT System, Abbreviated)

Phylum	Subphyla	Classes	Orders	Suborders	Families
	Type of Genetic Material	Symmetry of the Nucleocapsid Helical (H) Cubical (C) Binal (B)	Nucleocapsids Naked (N) Enveloped (E)	Rigid (R) Flexuous (F)	For Helical Viruses: Diameter of Nucleocapsid For Cubical Viruses: Number of Triangulation Number of Capsomeres
	DNA: Deoxyvira	H. Deoxyhelica	E: Chitovirales		100 Å: Poxviridae (vaccinia, smallpox)
		C: Deoxycubica	N: Haplovirales		1–12: Microviridae 3–32: Parvoviridae (defective adeno-associated) 7–72: Papilloviridae (papilloma, warts) 25–252: Adenoviridae (adenoviruses) 81–812: Iridoviridae ?: Inoviridae
			E: Peplovirales		16–162: Herpesviridae (herpes simplex, chicken pox, infectious mono)
		B: Deoxybinala	N: Urovirales		Phagoviridae (some phages; T series of *E. coli*) —

Vira					
	RNA: Ribovira	H: Ribohelica	N: Rhabdovirales	R: Rigidoviridales	102–130 Å: Dolichoviridae 150: Protoviridae 200: Pachyviridae
				F: Flexiviridales	100–110: Leptoviridae 120–130: Mesoviridae 150: Adroviridae
			E: Sagovirales		90 Å: Myxoviridae (influenza) 180: Paramyxovirida (parainfluenza, mumps, measles, distemper) Stomatoviridae
		C: Ribocubica	N: Gymnovirales		3–32: Napoviridae (polio, mengo, colds) 9–92: Reoviridae (blue tongue)
			E: Togavirales		?: Arboviridae (rubella, yellow fever, hog cholera, some arboviruses)

ing single-stranded DNA have also been described. The presence of an unusual base, 5-hydroxymethylcytosine, in the DNA of the T-even phages was very useful to those who first studied the intracellular events of phage development.

SIZE MEASUREMENT

There are four methods used to measure the size of viruses.

Microscopic examination. Size can be estimated by direct examination in the transmission electron microscope.

Filtrability. In this method the virus preparation is passed through membrane filters of known pore size. The filtrate is then assayed for infectivity. The size of the virus is larger than the last filter that it did not pass through, and smaller than the ones through which it passed.

Centrifugation. Virus sizes can also be estimated by the rate at which they settle in a centrifuge. A problem with this method is that shape as well as size affects the sedimentation rate.

Radiation. X rays or other high-energy radiation can also be used to estimate the size of viruses. The size of the virus can be calculated by considering each virus particle as a target, and the radiation needed to inactivate two-thirds of the viruses.

ENUMERATION (QUANTIFICATION)

There are a number of ways to estimate or assay the number of viruses present in a culture or in a fluid such as a sewer effluent. The most widely used method for the assay of infectious viruses is the *plaque assay*. If a fluid containing phages is being assayed, a certain dilution is inoculated onto a petri plate uniformly covered with a layer (lawn) of susceptible cells. Some types of viruses can be assayed on continuous single-cell sheets (monolayers) of tissue cultures. After a day or so, a clear area or plaque appears in the bacterial lawn or monolayer. Under ideal conditions, each plaque is started from a single infectious unit (PFU, plaque-forming unit). The number of viruses in the original fluid can then be calculated.

PURIFICATION

Viruses can be isolated from large volumes of microbial or tissue culture media, from body fluids, and from infected cells. Ammonium sulfate, ethanol, or polyethylene glycol can be used to precipitate the viruses. They can then be separated from host materials by passage through fine filters or separated and concentrated by differential centrifugation, density gradient centrifugation, column chromatography, or electrophoresis.

CULTIVATION

Phages are, of course, cultured in appropriate strains and species of bacteria. Some viruses seem to grow only in specific types of organisms. An increasing number of viruses can be grown in tissue cultures. *Primary tissue cultures* are made by dispersing cells (usually with the enzyme trypsin) from host tissues. Some diploid *secondary* cultures can be transferred serially for 40 to 50 passages. Some cell strains with altered and irregular numbers of chromosomes can be transferred perhaps indefinitely. Viruses are restricted to the types of tissue cultures in which they are capable of replicating. Many viruses are also grown in chick embryos.

VIRUS REPLICATION

Virus replication is conveniently divided into seven steps: attachment (adsorption), penetration, transcription, replication of nucleic acid, synthesis of other viral constituents, assembly, and liberation (release).

Attachment (Adsorption). The proper match between specific cell surface components and particular viruses is a requirement for interaction and attachment of viruses. Receptor sites are on the normal outer layer of the host cell, which may be composed of proteins, polysaccharides, or lipoprotein-polysaccharide complexes. Rate of viral attachment is determined by the type of virus, concentration of host cells, virus specific ion concentrations, and temperature.

Penetration. The mode of penetration varies among different types of viruses. Penetration of the T_4 bacteriophage of *E. coli* is a multistep process. After the tail fibers of the phage attach to the surface of the bacterium, they contract and bring the core of the tail in contact with the surface. Next an enzyme at the tip of the tail digests a small hole in the cell envelope, and the tail sheath contracts, injecting the DNA from the head through the tail and into the bacterium. The head and tail remain on the surface. Animal virions enter cells by pinocytosis or phagocytosis.

Transcription. Once viral nucleic acid reaches compatible host cell cytoplasm, part of it is transcribed by host cell RNA polymerase to synthesize "early" viral messenger RNA (mRNA). Host cell ribosomes are used to translate the mRNA into the enzymes required for the replication of the viral nucleic acid. Other early enzymes that may be formed include those involved in the synthesis of viral-specific nucleic acids (e.g., 5-hydroxylmethylcytidylic acid) or deoxyribonucleases which destroy all or some of the host DNA.

Replication of Nucleic Acid. Since the nucleic acids of viruses vary in terms of type (RNA or DNA) and strandedness (single or double), the pattern followed during replication necessarily also varies.

IN DOUBLE-STRANDED DNA. The replication of nucleic acid in phages with double-stranded DNA is similar to replication in bacteria. Two modes are possible, the *symmetric mode,* discussed on page 101, and the *asymmetric,* or *rolling circle mode,* as in E. coli. A newly synthesized virus-specific DNA polymerase is often involved in the DNA synthesis.

IN SINGLE-STRANDED DNA. In the case of a single-stranded DNA virus, such as phage ϕX 174, the first step in replication is the synthesis of a complementary strand so that it becomes a temporary double-stranded *replicative form.* DNA synthesis then proceeds as in normally double-stranded forms. After replication the complementary strand is discarded.

IN DOUBLE-STRANDED RNA. Replication of viruses with double-stranded RNA is basically similar to that in double-stranded DNA viruses, except that different nucleotides are involved.

IN SINGLE-STRANDED RNA. Two mechanisms for nucleotide replication have been found in viruses with single-stranded RNA. In one type, which includes RNA phages and polio virus, the viral RNA serves as a messenger for the synthesis of *RNA replicase,* an enzyme that makes a complementary strand of RNA. The two strands remain associated during replication. In another type, found in certain tumor viruses, replication takes place by *reverse transcription.* Here the viral RNA serves as a template for the synthesis of a complementary strand of *DNA.* A second strand of DNA complementary to the first is synthesized; the resulting double-stranded DNA then becomes integrated into the host genome. The host cells transform into tumor cells and the virus-derived DNA is replicated and transmitted to daughter cells at each division. Under certain conditions, single-stranded RNA synthesis is resumed, leading to the formation of new infective virions.

Synthesis of Other Viral Constituents. After the beginning of DNA replication, "late" mRNA codes for a second set of viral proteins required for the synthesis of the capsomeres. Soon after, these structural proteins can be detected in the cell.

Assembly (Maturation). The protein capsomeres spontaneously assemble into the capsid. In the case of binal phages with several different kinds of proteins, the head forms first. The tail core and endplate come together and acquire a sheath. Heads then join the completed tails. Tail fibers are added last. Details of the various steps in protein synthesis and self-assembly have been clarified through the study of mutant viruses blocked at various stages in the process. *E. coli* phages T4 and T7 are among the best-known viruses used in this work.

Liberation. As phages are assembling, another viral late protein, lysozyme, is formed in the cell. This enzyme hydrolyzes the linkages between the sugar residues in the peptidoglycan layer of the host cell until the wall becomes so weak it ruptures, liberating the viruses. Some phages, such as fd, a filamentous DNA phage, are liberated by extrusion from the surface of the host. Animal viruses are also extruded from the surfaces of host cells. Their peplos, or mantle, is acquired as they bud off from the cell nucleus or the cell membrane. Before this happens, virus-specific antigens become identifiable on the cell surface.

VIRUS RESTRICTION AND MODIFICATION

One form of host resistance to viral attack is the production of *restriction enzymes* that cleave viral DNA at one or several places to prevent replication. Restriction enzymes are very specialized DNAases which attack double-stranded DNA only at sequences that have two-fold symmetry around a particular point. This ability to cleave DNA at particular points into smaller fragments with defined termini makes restriction enzymes valuable tools for DNA research. Hosts protect their own DNA from restriction enzymes by methylation of specific bases within the recognition sequence. This protection is brought about by *modification enzymes* specific for the hosts' restriction enzymes. Only viruses that do not have nucleic acid sequences attacked by the host's particular restriction enzymes are able to initiate replication.

LYSOGENY

In lysogeny, another virus-host relationship, a virus penetrates a host cell but does not destroy the host's DNA, nor does it undergo vegetative replication. Instead it enters a relatively stable genetic relationship with the host. Sometimes observable genetic changes are brought about by lysogenization. A lysogenic phage, β, converts nontoxin-producing pathogenic strains of *Corynebacterium diphtheriae* into toxin-producing strains (*phage conversion, lysogenic conversion*). Two modes of integration are known among phages, one typified by the *E. coli* phage λ which becomes physically integrated into the hosts' chromosome at a specific site, and the other typified by phage Pl which becomes a separate, circular *prophage* (*plasmid*). In either case, the repressor molecules do not interfere with the replication of prophage in synchrony with the host cells' replication. A number of mutagenic agents such as ultraviolet light, x rays, and nitrogen mustards at sublethal doses can *induce* or cause host systems with integrated (latent, temperate) prophages to return to the *lytic* or actively replicative state. In the case of λ, it is known that after treatment, repressor protein is inactivated, allowing transcription of an early viral gene which shuts off the mRNA coding involved in repressor protein formation. Another early viral gene causes large portions of the λ DNA to be transcribed. This replication of λ DNA takes place by the *rolling circle mode of replication* (Fig. 6.1), which proceeds in only one direction and produces very long chains of DNA. The long DNA concatamers are then cut into virus-sized lengths by a cutting enzyme.

TUMOR VIRUSES

Although it was suggested more than 70 years ago that certain cancers were caused by viruses, the idea was not generally accepted by the scientific community until many of the molecular genetic details of bacteriophage lysogeny were worked out. Tumor-producing (*oncogenic*) viruses are found in several groups of DNA viruses (adeno-, herpe-, pox-, and papovaviruses) and a limited number of RNA viruses (leukovirus group: leukemia, lymphoma, and sarcoma). The invasion of an oncogenic virus into susceptible cells causes a number of changes in the cells. The process of change is called *transformation,* an unfortunate term, since the same word is also used by microbiologists to describe a totally unrelated genetic phenomenon. One of the best-known changes can be seen in tissue culture cells. Normal cells exhibit *contact inhibition,* a phenomenon in which contact between cells inhibits both movement and cell division. Transformed cells, however, continue to divide and form tumorlike cell masses. Even though infectious viruses cannot be detected in transformed tissue culture cells, it has been shown that the viral DNA (or DNA complementary to tumor virus RNA formed by reverse transcriptase) is integrated into the DNA of the host and is reproduced at each cell division.

Among the best-known research in the long history of oncogenic virus is that of P. Rous, who showed that chicken leukemia could be

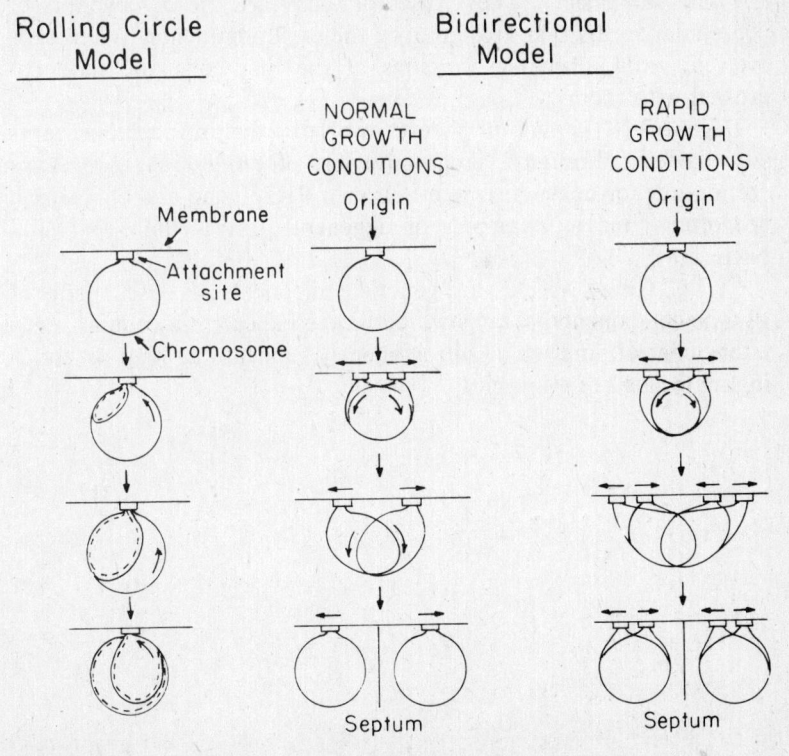

Fig. 6.1. Rolling circle and bidirectional models of DNA replication.

transferred by means of cell extracts; that of R. Shope, who demonstrated the tumor-inducing capacity of cell-free filtrates of a rabbit papilloma (wartlike tumor); that of J. Bittner, who discovered a virus in mouse milk which could induce mammary cancer; and that of H. M. Temin and D. Baltimore, who discovered the reverse transcriptases of RNA tumor viruses. The widespread occurrence of virus-induced tumors in animals has intensified research on human cancer viruses. Evidence seems convincing that a herpes virus, the Epstein-Barr (EB) virus, is the causative agent of Burkitt's lymphoma.

ORIGIN AND EVOLUTION OF VIRUSES

Three general hypotheses have been proposed to explain the origin and evolution of viruses.

Precelluar Theory. The viruses of today are the descendants of precellular forms that arose during the evolution of life. When cells evolved, viruses became parasites of the first cells and have coevolved with them.

Degenerative Evolution. Viruses evolved from intracellular parasitic bacteria through a progressive loss of cellular characteristics (retrograde, or degenerative, evolution). Rickettsiae and chlamydae are often cited as examples of degenerate intracellular parasitic bacteria.

"Wild" Genes. Viruses are genetic components, or descendants of genetic components, of cells which have escaped from cellular regulatory control, and which autonomously replicate as long as building materials are available.

Chapter 7

GENETICS OF MICROORGANISMS

PROCARYOTES AND VIRUSES

The small size and rapid generation times of microorganisms make them ideal subjects for certain types of genetic research. The enormous strides in molecular genetics in recent years were made possible by dedicated pioneers who recognized microbial potential and exploited them as tools for this exciting research. Current molecular biological research (including genetic engineering) suggests that we are on the verge of significant breakthroughs that will benefit our health and welfare in the years ahead.

Milestones.

Luria and Delbruck—1943. Performed a classical experiment which proved that bacteria have stable hereditary mechanisms. In their carefully designed study, they showed that phage-resistant bacteria arose spontaneously by mutation independent of the selective agent, the phage, used to demonstrate the phenomenon.

Avery, MacLeod, and McCarty—1944. Demonstrated that DNA was the hereditary agent in the transformation of *Pneumococcus* types.

Beadle and Tatum—1958. Won the Nobel Prize for their pioneering genetic studies with nutritional mutants (auxotrophic strains) of the haploid pink mold *Neurospora crassa*. They showed that each mutation was the result of the alteration in a singe gene and developed a conceptualization of gene function known popularly as the "one gene-one enzyme" hypothesis.

Watson, Crick, and Wilkins—1953. Won the Nobel Prize for their fundamental contributions on the molecular structure of DNA and how it might be replicated.

Lederberg and Tatum—1946. First convincing demonstration of conjugation and recombination in bacteria.

Hayes—1952. First evidence of sexual differentiation in *E. coli*.

Zinder and Lederberg—1952. First demonstration of transduction in bacteria.

Lwoff—1953. Formulated a model to explain the phenomenon of lysogeny.

Jacob and Monod—1961. Developed the operon conceptual model of genetic control of enzyme synthesis.

Holley, Khorana, and Nirenberg—1968. Shared the Nobel Prize for their contributions to the understanding of the genetic code and its function in protein synthesis.

The Genetic Code. Genes are made up of long, continuous double helices of DNA. Sequences of three adjacent nucleotide bases, *codons*, code for each of the 20 amino acids commonly found in proteins. There is some redundancy in the code so that some individual amino acids (serine, leucine, arginine) are coded by as many as six codons and some (tryptophan, methionine) with as few as one. The majority have two (phenylalanine, cysteine, tyrosine, histidine, glutamine, aspartic, glutamic, lysine, asparagine) or four codons (proline, valine, alanine, glycine, threonine). One amino acid, isoleucine, has three codons. The mRNA triplets for the same amino acid are usually closely related (e.g., CGA, CGG, CGC, and CGU all code for alanine). Note it is the last of the three nucleotides in the sequence which changes. This aspect of the code where only two of the three bases in the code may be significant for certain amino acids is known as degeneracy. As mentioned previously (Chapter 5), the first step in the process of protein synthesis is the transcription of the base sequences of DNA into complementary strands of mRNA which in turn become translated into amino acid sequences in proteins. A few mRNA triplets (UGA, UAG, UAA) called *nonsense codons* do not code for any amino acids. Instead they signal the end of the genetic coding for a specific protein (punctuation). Several triplets (GUG, AUG) which are the codons for N-formyl methionine at the beginning of mRNA serve as signals to initiate the start of translation of a polypeptide chain.

A rare, but interesting, translation phenomenon, that of *overlapping genes*, has been found in the small bacterial virus ϕX 174 which has only seven genes. By beginning transcription at different sites out of phase with each other, the same small segment of DNA codes for more than one protein.

Mutations. Mutations in DNA are rare events occurring at rates so small (1×10^{-6} to 1×10^{-10}) that only one bacterium in a million to tens of billions is likely to undergo a mutation of a given kind. At the molecular level, mutations are changes in the nucleotide sequences of DNA. These changes modify the information and result in the formation of altered protein.

TYPES OF MUTATION. The changes in nucleotide sequences can occur by substitution of one nucleotide for another (point mutations), by deletion of one or more nucleotides, or by the insertion of additional nucleotides.

Point Mutations. Among point mutations, the substitution of one purine for another or of one pyrimidine for another is known as a *transition* while the substitution of a purine for a pyrimidine or the reverse is known as a *transversion.* Point mutations can result in coding for a different amino acid, no change in the amino acid specified because of degeneracy in the code (*silent mutations*), or formation of a *nonsense codon* which would terminate translation.

Deletion and Insertion Mutations. Deletion or insertion of nucleotides can cause *frame shifts,* in which adjacent nucleotides following the deletion/insertion become grouped into different triplet sequences (Table 7.1). Generally frame shifts result in the formation of nonfunctional proteins.

Table 7.1 Mutations in the Genetic Code

Normal reading frame	CAN	THE	MAN	PAT	HER	CAT	
Point mutation	CAN	THE	MAN	PAT	HER	RAT ↑	
Frame shift deletion	CAT ↑ N. deleted	HEM	ANP	ATH	ERC	AT	
Frame shift insertion	CAN	THE	MAN	PAN ↑ N inserted	THE	RCA	T

Benzer, in his now classical fine structural mapping studies of mutations within 300 distinct sites of the r II gene of phage T4, has shown still another interesting aspect of mutation. He observed two "hot spots," sites where the frequency of mutation was, respectively, 50 and 275 times higher than at other sites.

MUTAGENIC AGENTS. Chemical or physical agents such as x-rays, ultraviolet light, nitrous acid, base analogs (2-aminopurine, 5-bromouracil), and aklylating agents (nitrosoguanidine, proflavine) which affect the structure of DNA increase mutation rate and are known as *mutagens*. A number of cellular systems to repair DNA damage have been described. They are activated when DNA damage blocks replications. Some enzymes recognize and excise the incorrect complementary sequence using the undamaged strand as a template. There are also DNA polymerases capable of adding bases to a single-strand end of DNA without the aid of template instruction.

Recombination. Recombination is the process by which progeny are produced whose genotypes are different from those of the parents. Recombination in procaryotes differs from sexual phenomena in eucaryotes in that a true zygote is not formed and only a fraction of genetic material from the donor cell (*exogenote*) is transmitted to a recipient (*endogenote*). These partial zygotes are called *merozygotes*. If a sequence of base pairs in the exogenote is homologous with the base pairs in a segment of the endogenote, all or part of it pairs and becomes integrated into the endogenote. If the exogenote does not pair, it may persist as a separate genetic entity and if it carries the necessary genetic elements it may replicate separately (*replicon*). Unintegrated exogenotes may be enzymatically degraded by endonucleases known as host restriction enzymes. Some mutant strains, such as rec$^-$ (recombination minus), which are unusually sensitive to ultraviolet radiation and deficient in dark DNA repair enzymes, are unable to form recombinants.

In procaryotes, three mechanisms are involved in recombination: conjugation, transformation, and transduction. Although at one time it was easy to draw clear distinctions between them, recent work has tended to blur the boundaries.

CONJUGATION. Conjugation, the transfer of a strand of DNA from one cell to another, requires direct contact between the cells. One cell acts as genetic donor (male) and the other as genetic recipient (female). Maleness is determined by a transmissible, autonomous genetic element (*plasmid*) known as the F (fertility) factor. (Plasmids were discussed in Chapter 3.) The F factor may be integrated into the male chromosome (Hfr) or exist in the cytoplasm as an independent, double-stranded, continuous loop of DNA (F$^+$). The site for chromosomal integration of the F factor varies with different strains.

On the surfaces of cells that carry the F factor are structures called *sex pili*. A recipient cell makes contact with a sex pilus; the pilus retracts rapidly, bringing the cells close together. Contact triggers activity in the F plasmid. One strand is broken at the replication origin and enters the recipient cell beginning with the 5' end. Through the action of DNA polymerase, complementary strands are synthesized in both the donor and the recipient. The replication origin is in the middle of the F factor so that in Hfr (high-frequency recombination) strains, the exogenote chromosome is also transferred. Within 25 minutes after contact, a strand of the chromosome of almost every mated Hfr in a population is in the process of being slowly injected into the F$^-$ cell. If conjugation is not interrupted, a strand of the entire donor chromosome representing 5×10^6 base pairs can be injected in about 90 minutes. Complete transfer of the entire chromosome, however, is a rare event, taking place in only one in 10,000 matings. Those characteristics closest to the point of integration of the F factor on the chromosome have the highest possibilities of transfer. There is a 60 percent chance for recombination for factors that enter the recipient cells within the first 7 minutes, but only a 40 percent chance for factors that take 12 minutes. At 17 minutes the chance for recombination is only 15 percent. Although Hfr strains transmit chromosomal genes at high frequency, the offspring of their matings are rarely F$^+$ or Hfr because the entire F factor is rarely transferred, since part of it is at the terminal end of the chromosome. The low frequency of recombinant cells (1×10^{-5}) in F$^+$ $\times$ F$^-$ crosses has been shown to be the result of the rare conversions of F$^+$ to Hfr cells.

TRANSFORMATION. Transformation is genetic recombination brought about by the addition of DNA extracted from a particular strain of cells which is picked up and integrated into the genome of a related recipient cell. Transformation is one of the earliest phenomena of bacterial genetics to be studied in the laboratory (Avery, MacLeod, and McCarty, 1944). The DNA of donor cells, extracted and purified with appropriate techniques that protect and preserve it, is added to cultures of genetically related potential recipient cells. Only certain strains (*competent*) are capable of binding the DNA fragments and integrating short segments containing a few genes (~10,000 nucleotide pairs) into their chromosomes. In addition to *Streptococcus pneumoniae*, the organism used in the first experiments, there are competent strains of *E. coli* and species of *Bacillus, Hemophilis, Neisseria,* and *Pseudomonas.*

TRANSDUCTION. Transduction is a widespread bacterial genetic phenomenon in which host DNA is transferred from one cell to another by temperate viruses. Two types are recognized: specialized and generalized.

Specialized Transduction. Here a restricted group of host genes is integrated directly into the viral genome, usually replacing virus genes. The best studied specialized transduction system is the lambda (λ) phage of *E. coli*. This phage becomes integrated into the host DNA adjacent to the cluster of genes that control the enzymes involved in galactose utilization. When the galactose cluster is inserted into the phage genome in place of phage genes, the phage becomes defective (λdg) and can no longer make mature phages. However, if another phage, called a *helper,* is used along with λdg in a mixed infection, the defective phages are replicated and transduce the galactose genes. Lysogenic conversion (p. 100) is really a specialized transduction.

Generalized Transduction. In this rarer genetic phenomenon, virtually any genetic marker can be carried from donor to recipient by temperate phages. During the lytic cycle, host DNA is broken down into smaller pieces, some of which become incorporated into the virus genome to be carried to their new hosts. For any given genetic marker, the probability of being carried in generalized transduction is quite rare (1×10^{-6} to 1×10^{-8}). Some exceptional phages (e.g., Coli Phage Pl) can transduce markers at higher frequencies ($\sim 1 \times 10^{-5}$).

Genetic Mapping of E. Coli. By the use of various Hfr strains, each with different insertion sites on the chromosome, the arrangement and orientation of a large number of chromosomal genes can be determined. The overall locations of various genes on the chromosome are mapped by means of interrupted matings. Map distances are calculated in number of minute differences between genetic markers. Transduction is more useful for determining the order and location of closely linked genes and gene clusters. The *genome* (complete set of genes) of *E. coli* contains around 4×10^6 base pairs. Genes that have been mapped are not randomly distributed. Mapping the genes in metabolic pathways has shown that genes in the same pathway are often closely linked or clustered. In pathways with clustered genes, a single mRNA is transcribed which then directs the synthesis of all the enzymes in the cluster.

Operon Concept. From their studies of the genetic control of lac-

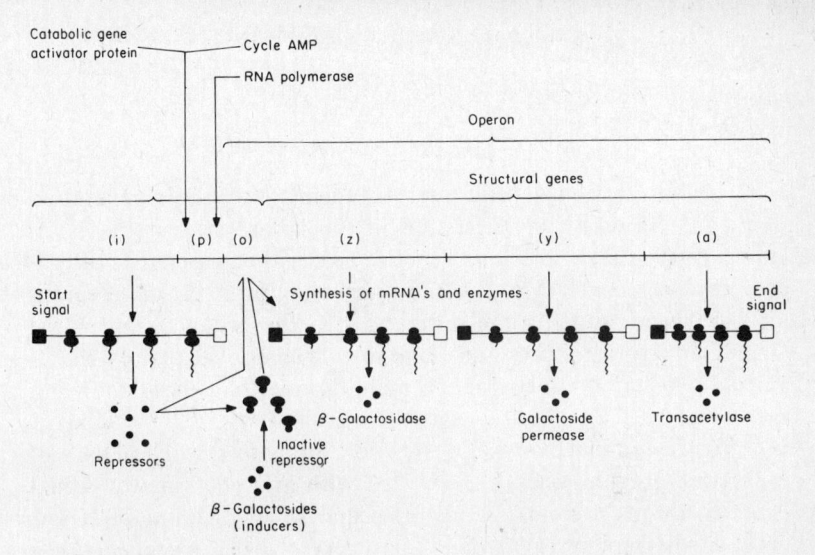

Fig. 7.1. Operon model.

tose metabolism Jacob and Monod developed their comprehensive model (operon) (Fig. 7.1) to explain the genetic control of clusters of structural genes governing the synthesis of enzymes of a particular metabolic pathway. When lactose or closely related inducers are added to a culture, the syntheses of β-galactosidase, galactoside permease, and thiogalactoside transacetylase are increased (z, y, a in Fig. 7.1). Rates of synthesis are controlled by three types of regular genes, repressor, promoter, and operator. The repressor gene codes for a repressor protein which binds to the DNA of the operator gene, preventing its transcription. The promoter gene is the site on the DNA where the RNA polymerase binds to initiate mRNA synthesis.

Recombinant DNA and Genetic Engineering. It is now technically feasible, with the aid of restriction endonucleases, DNA ligases, and DNA polymerases, to dissect and remove DNA segments from viruses and other organisms and incorporate them into plasmids. Since plasmids can be independently replicated and are easily isolated from cells, significantly large amounts of DNA that carry genes of interest can be obtained. Recombinant DNA techniques have the potential for producing bacterial strains that have special utility. An example is the insulin-synthesizing bacterial strain recently constructed. Because the possibility exists of fabricating a "doomsday bug," guidelines have been established for recombinant DNA research.

EUCARYOTIC MICROORGANISMS

Although most eucaryotic microorganisms have short generation times and are easily manipulated in the laboratory, much less is known about their molecular genetics than is known of that of procaryotes. In recent years, however, increasing numbers of researchers have been studying the differences in the organization of the genetic apparatus of pro- and eucaryotes. Genetic regulatory mechanisms in eucaryotes seem much more diverse than those known in the procaryotes.

Organization of DNA in Eucaryotic Chromosomes. The DNA of eucaryotic chromosomes is associated with an approximately equal mass of fibrous proteins (*chromatin*) and small amounts of RNA. Most of the protein is *histone,* a category of basic proteins with a high proportion of arginine and lysine. The amount of nonhistone protein in chromosomes varies widely but is small in relation to histones. The various histones in the chromosomes interact with each other and with DNA to form subunits known as *nucleosomes* (ν[nu] bodies, PS particles). A typical gene has three or four nucleosomes, each around 7.0 to 10 nm in diameter with about 150 base pairs of DNA within each particle and another 30 to 60 base pairs as a flexible strand between each particle. Nucleosomes are packed into higher orders of structure. The structure of histones can be modified reversibly by the addition or removal of phosphate or acetate groups and by the oxidation and reduction of sulfhydryl groups.

Regulatory Mechanisms for Gene Expression. There are two possible regulatory mechanisms for gene expression in eucaryotes: folding and unfolding of the chromosome and gene regulation.

HISTONE MODIFICATION. Several cyclic AMP (cAMP) dependent histone kinases have been isolated which could mediate the process of folding and unfolding of the chromosomes. Some evidence suggests that histone modification and the resulting unfolding of chromatin to expose DNA is a regulatory mechanism to turn blocks of genes off and on. Nonhistone chromosomal proteins (NHC proteins) have the capacity to bind to regulator sites of specific genes and may alter or displace histones so that the genes become available for transcription.

GENE REPLICATION. Gene replication seems to be another regulatory mechanism for the expression of certain genes in eucaryotes. Multiple copies of specific genes in the nucleoli of amphibian oocytes enhance the synthesis of rRNA and tRNA.

Mitochondria and Chloroplasts. In chapter 3, we discussed the evidence that mitochondria and chloroplasts may have originated as procaryotic symbionts in ancestral eucaryotic cells. The DNA of mitochondria and chloroplasts can be isolated separately from nuclear DNA and characterized. In common with procaryotic DNA, the DNA of mitochondria is not associated with protein and has a double-stranded circular structure. The mitochondria of some animal cells contain as many as six loops of DNA. The nucleotide base composition of chloroplast DNA is higher than that from the mitochrondria of the same cell. Estimates from studies of *Chlamydomonas* suggest that the DNA codes for approximately 180 genes.

When mitochondria and chloroplasts are isolated *in vitro*, they are capable of protein synthesis. They have their own ribosomes, transfer RNA, DNA-dependent RNA polymerases, and DNA polymerases. Like protein synthesis by procaryotic ribosomes, protein synthesis by mitochondrial and chloroplast ribosomes can be inhibited by chloramphenicol and erythromycin, but not by cycloheximide (an inhibitor of eucaryotic protein synthesis).

Mitosis and Meiosis. While the processes of mitosis (nuclear division) and meiosis (reduction division) in eucaryotes are familiar to every biology student, a number of eucaryotic microorganisms have some unusual mechanisms for accomplishing these functions. Among the protozoa there is a great variety of mitotic types. In certain groups, which may be more primitive, the nuclear membrane does not break down during mitosis; the chromosomes attach themselves to the inside of the nuclear membrane and separate as the membrane grows apart and eventually constricts. The similarity of this chromosome type of separation in mitosis and the separation of daughter chromosomes in bacteria has not gone unnoticed. The complicated reduction division of some postconjugant molds (*Aspergillus*) will be discussed in greater detail below.

Eucaryotic Microbial Genetic Systems of Special Interest

SACCHAROMYCES CEREVISIAE. Mitochondrial genetics is best known from studies of respiratory-deficient (RD) mutant strains of yeasts. RD mutants have lost their ability to respire and must obtain all their energy from fermentation reactions. Such mutants are thought to have a disrupted ultrastructure that renders them incapable of synthesizing respiratory chain enzymes. Mutations in the nuclear genes of yeasts also have been detected; these mutations affect the synthesis of cytochromes and other mitochondrial proteins. The current interpretation of the evidence is that genetic systems in the mitochondria and the nucleus work in concert to produce the final phenotype of the organelle. Antibiotic-resistant mutant haploid strains of yeasts have been used to show recombination in mitochondrial genes contributed by different parental strains and to construct a partial linkage map of the mitochondrial genome.

CHLAMYDOMONAS REINHARDI. Various types of mutants have been used to study chloroplast genetics in the unicellular green alga *Chlamydomonas reinhardi*. Cells of this species are haploid and are differentiated into two mating types, mt^+ and mt^-, determined by a single gene. Cells of opposite mating types fuse to form a zygote which matures and by meiosis produces four haploid offspring. Using chloroplast genetic markers for antibiotic resistance, photosynthesis deficiency, and temperature sensitivity, it has been shown that almost all the postzygotic progeny inherit their chloroplasts from the mt^+ parent. The chloroplast DNA from the mt^- parent is usually degraded. In less than 1% of the crosses, degradation does not happen and the offspring have heterozygous chloroplast DNA. Ultraviolet irradiation of the mt^+ parent prior to mating increased the percentage of chloroplast heterozygotes and made it possible to construct a genetic linkage map of the chloroplast chromosome.

NEUROSPORA CRASSA. This ascomycetous fungus has been used to study inheritance of both nuclear and extranuclear genes. It produces asexually two types of conidia, or fruiting bodies: multinucleate *macroconidia* and uninucleate *microconidia*. These fruiting bodies can grow into new mycelia or can fertilize another specialized reproductive cell, the *ascogonium* (♀). The zygote undergoes meiosis and produces eight offspring, called *ascospores*. The eight ascospores are aligned in a single row in such a way that it is possible to identify and separate spores that are the products of each of the steps in meiosis (eight spores are formed from the two meiotic divisions and one mitotic division). This characteristic facilitates the study of crossing over. The mitochondrial mutants (mi) of this mold have an inheritance pattern similar to that of the chloroplasts of the alga *C. reinhardi*. Crosses between strains of this mold inherit their mitochondrial characteristics from the ascogonium (♀ parent).

Another type of genetic recombination, *heterokaryosis*, has been observed in *N. crassa*, many other molds, and in *Streptomyces*, an important antibiotic-producing genus of moldlike procaryotes (Actinomycetales). Heterokaryosis literally means more than one kind of nucleus. It is brought about after the contact and fusion of two vegetative hyphae. The nuclei of both parental types are maintained as separate genetic entities in the mycelium. Depending upon environmental conditions, both types of parental nuclei reproduce as the mycelium grows. The nuclei segregate as separate genetic entities when conidia are formed. In heterocaryons, mitochondrial inheritance is not linked to nuclear inheritance; the mitochondria in the conidia can come from either parental stock.

ASPERGILLUS NIDULANS. Heterokaryosis and the phenomenon of parasexuality have been studied more intensively in the ascomycete mold, *Aspergillus nidulans*. As in *Neurospora crassa*, hyphal fusion leads to heterocaryons with mixed phenotypic characteristics but with nuclei which do not fuse but rather multiply by mitosis independently of one another. Rarely (1×10^{-7}) nuclear fusion occurs, leading to the formation of a heterozygous diploid zygote. Diploid spores can be recognized and can be used to grow diploid mycelia. Haploidy is eventually restored to the progeny of diploid strains by the occasional loss of one member of each chromosome pair during mitosis. Since individual chromosomes can be derived from either parent, the resulting haploids are recombinants with some characteristics from both. The phenomenon is considered parasexual because it does not involve the more common processes of gametic fusion and meiosis.

Chapter 8

SYMBIOSIS

The interactions between different types of organisms living together for their mutual benefit have long fascinated microbiologists. It is hard to imagine what world ecosystems would be like in the absence of many of the better known symbiotic systems. Paleontologists suggest that symbiotic associations have been the driving force, or the keys to the success, of many groups during their evolution.

SOME TERMINOLOGY

Although the term symbiosis was first coined to mean any close relationship between two organisms, many workers have tended to apply it only to associations that are beneficial to both partners. Common usage has also tended to call the larger organism in the association the host and the smaller one the symbiont, or parasite. Strictly speaking, symbiosis is a general term for intimate associations between different types of organisms. Its two main functions are protection and the fulfilling of nutritional needs, either directly or by providing a favorable position for nutrient uptake.

Types of Symbiosis. Symbiosis has been classified according to degree of benefit and physical arrangement of the association.

DEGREE OF BENEFIT. The four main categories of symbiosis present a spectrum of species interactions from mutually beneficial to harmful.

Mutualism. Both organisms benefit.
Commensalism. One organisms benefits, the other is unaffected.
Neutralism. Neither organism benefits or is harmed.
Parasitism. One organism benefits, the other is harmed.

PHYSICAL ARRANGEMENT OF THE ASSOCIATION. Symbionts are often categorized according to the intimacy of the association.
Ectosymbiosis. One organism lives on the outside of the host with little penetration.
Endosymbiosis. One organism lives inside the body cavity, the tissues, and sometimes even the cells of the host. In those environments it is provided with conditions that favor its survival, growth, and reproduction.
Gnotobiology. This is a subdiscipline of the sciences of microbiology and nutrition. Using aseptic and other appropriate techniques, interactions between organisms are studied under conditions in which all the organisms in the system are known. For example, to test the interrelation of certain seeds and soil microorganisms, surface-sterilized seeds are planted in sterilized soils and grown in *axenic* (germ-free) environments in which all air and water entering the system is sterilized. Axenic soil microorganisms are then added to the plants in some chambers but not to those in others (the controls). Similar experiments are conducted with gnotobiotic animals. To study the effects of intestinal microflora on growth, for example, the animals to be used in the experiment are delivered aseptically by cesarean section and raised in axenic environments.

ASSOCIATIONS BETWEEN PHOTOSYNTHETIC AND NONPHOTOSYNTHETIC PARTNERS

In such associations, the photosynthetic partner provides carbohydrates by using light energy to fix carbon. The other partner serves a variety of functions. In some of the invertebrate symbioses the photosynthetic partner also enhances skeletal building.

Mycorrhiza: Fungi and Higher Plants. Soils abound with microorganisms, some of which enter into mutualistic symbioses with other organisms. Most higher plants have root systems that enter into associations with a number of species of soil fungi. Two general categories of mycorrhiza are recognized: *ectomycorrhiza* in which the fungi form an extensive network on the outside of the roots with only little penetration of the hyphae into the roots themselves; and *endomycorrhiza* in which most of the fungal mycelium is embedded within the roots. The relationship is initiated by the attraction of fungi to organic compounds released into the soil by the plant roots. After growing toward the root, the fungi invade the roots themselves. Depending upon the two species involved, the fungal mycelium penetrates individual root cells with projections known as *haustoria,* or grows intracellularly in characteristic branching networks called *arbuscules*. Gnotobiotic experiments have shown that both partners benefit from the relationship. The fungi generally use nutrients such as carbohydrates manufactured by the plants. The uptake of minerals such as K, PO_4^{3-}, NO_3^-, and water into the plant is greatly facilitated in plants with a mycorrhiza.

Bacteria and Root Nodules: Leguminous Plants (Nitrogen Fixation). The formation of root nodules in leguminous plants is perhaps the best-known example of the mutualistic interaction of soil microorganisms and higher plants. The bacteria "fix" nitrogen so it can be used by the plants; the plants provide nutrients needed by the bacteria.

The details of nodule formation differ among various legumes. The *preinfection* stage takes place in the soil in the immediate vicinity of plant roots. Among the low molecular weight metabolites that leak from plant roots is the amino acid tryptophan. Bacteria belonging to the genus *Rhizobium* pick up tryptophan and convert it to indole acetic acid (IAA), a plant hormone that causes the plant root hairs to curl around the bacteria as they grow. *Infection* begins if nitrate levels in the soil are low and if the strain of *Rhizobium* is matched to the particular legume. If these conditions are met, the root is induced to secrete an enzyme that digests the capsule of the bacteria which then becomes strongly bound to the root hairs. The bacteria enter the root hair tip and spread through it, forming an *infection thread*. Then they invade neighboring cells, destroying normal diploid cells until they reach tetraploid cells, which they stimulate to divide. This cell division produces the nodule. The infection thread in the tetraploid cells bulges out, forming vesicles that burst, releasing the bacteria. After the bacteria in the nodule multiply, they swell and become branched. These branched forms, called *bacteroids,* synthesize significant quantities of *nitrogenase*. The host cells also undergo morphological changes. Their chromatin fragments and their mitochondria swell, and large amounts of the O_2-binding protein *leghemoglobin* are synthesized.

It is presumed that the biochemistry of nitrogen fixation in leguminous root nodules is similar to that in free-living nitrogen-fixing procaryotes, which has been more extensively studied. Three coupled components of nitrogen-fixing systems are recognized: a hydrogen-donating and electron transport system (HDS), a nitrogen-activating system (NAS), and an ATP-generating system.

Lichens: Algae and Fungi. Many symbiotic associations permit their partners to survive and flourish under environmental conditions too harsh for most other organisms. In Antarctica, for instance, there are more than 350 species of lichens but only 2 species of flowering plants. Lichens are regular mutualistic associations between algae (cyanobacteria) and ascomycete fungi (basidiomycetes are found in some tropical lichens). After the symbiosis is established, the growth pattern leads to a composite morphology that can be identified without extensive laboratory bench work. The bulk of a lichen thallus is fungal. The algae are usually embedded in layers within a fungal matrix. In most associations, the algae are cupped around, or are penetrated by, special absorptive hyphae (*haustoria*). The penetration of the haustoria is through the cell wall only; electron micrographs show the cell membrane to be deformed into a sheath around each haustorium.

The green algal genera *Trebouxia* and *Trentepohlia* and the cyanobacterial genus *Nostoc* account for 90% of the known lichen mutualistic symbioses. Most lichens propagate by the liberation of *soredia*, hyphae containing a few algal cells. Another means of reproduction is by fungal spores, which germinate and link up with algal cells. Lichens have evolved the ability to withstand drought and scavenge minerals. The latter ability is due to the production and excretion of large quantities of characteristic organic lichen acids. Lichens grow slowly (2 to 4.5 cm/yr) but they live a long time. (Some thalli in the Alps are 600 to 1300 years old, and some from Greenland 1000 to 4500 years old.)

Radionuclide tracer experiments show that 20 to 50% of the carbon fixed photosynthetically by the alga moves from the alga to the fungus; the fungus converts it to forms of sugar alcohols (usually mannitol) which the alga cannot use. In return, the fungus provides a strong protective micro-environment in which the alga can absorb the water and inorganic nutrients it needs.

Algae and Aquatic Invertebrates. Endosymbiotic algae have been found in a great variety of marine and freshwater invertebrates, including protozoa, coelenterates, sponges, platyhelminths, annelids, tunicates, and mollusks. *Zoochlorellae* (green algal endosymbionts belonging to the Chlorophyta and Prasinophyta) are more common in fresh water invertebrates (e.g., *Paramecium bursaria,*

Chlorohydra viridis) while *zooxanthellae* (brown or olive brown endosymbionts belonging to the Pyrophyta and Bacillariophyta) are usually associated with marine invertebrates (e.g., corals and the giant tridacnid clams). *Cyanobacteria* and unicellular red algae have been found in a few cases of algal endosymbiosis. The most common zoochlorellae belong to the genus *Chlorella* and the most common zooxanthellae (found in some foraminifera, most corals, sea anemones, and tridacnid clams) is *Symbiodinium microadriaticum*. Radionuclide studies show transport of carbon fixed by the algae to the animal tissues.

NONPHOTOSYNTHETIC SYMBIOSES BETWEEN MICROORGANISMS AND METAZOANS

There are a number of mutualistic associations in which neither partner is photosynthetic. Such associations fulfill a variety of mutual needs and include symbioses between two microorganisms and between microorganisms and metazoans. The examples that follow are of the latter type.

Ruminant Symbiosis: Microorganisms and Mammals. Because of its three-dimensional cross-linking, cellulose is a particularly difficult structural polysaccharide for enzymes to digest. A battery of three enzymes is needed to release glucose units. Almost all the cellulose in the biosphere is degraded through the activities of microorganisms.

In the soils of forests and fields, and in the bottom of the sea, microbial degradation of cellulose takes place very slowly. However, during the course of evolution, many animals have developed internal organs that speed up the process. The best known of this type of organ is the rumen, which is really the analog of a giant fermentation chemostat. The volume of cow rumen is about 100 liters. Chewing and grinding is a very important part of the initial process of degrading the cellulose since it breaks cross-linking bonds and thus exposes increasing parts of the molecule to cellulase attack. After being finely ground, the food enters the rumen and is mixed with the resident microbial flora and protozoa ($\sim 10^{10}$ cells/ml), some of which bind themselves physically to the cellulose. The most common cellulose-digesting rumen bacteria belong to the genera *Bacteroides, Butyrivibrio, Ruminococcus,* and *Clostridium*. Most of the fermenting protozoa are members of an unusual group of ciliates, the Entodiniomorphida. They may serve yet another function in the rumen. Experimental evidence suggests that by eating large quantities of bacteria, the protozoa control bacterial densities and keep the bacteria actively dividing. Some of the important end products of rumen fermentation are acetate, formate, succinate, butyrate, isobutyrate, valerate, isovalerate, propionate, and lactate which are absorbed in the intestine.

Protozoa and Termites. The hind gut of termites and woodroaches is the site of their microbial cellulose fermentation system. More advanced termites do not have endosymbiotic bacteria in their hind guts, but instead have rich cellulose-digesting flagellated protozoan fauna. Evidence suggests that these flagellates, which are a morphologically unusual group of specialized protozoa belonging to the Hypermastigida and Trichomonadida, have coevolved with their hosts.

Luminous Bacteria and Fish. A number of unrelated species of fish and myopsid squids have specialized organs that contain luminous bacteria. The organs seem to serve as devices to promote schooling, mating, or the attraction of prey. Despite the diversity of fish that have luminous organs, all the bacterial endosymbionts that have been isolated in culture belong to species of the genus *Photobacterium*. The best known of these fish with luminous organs is *Photoblepharon*, which harbors luminous bacteria in special pouches under the eyes. In *Photoblepharon* the amount of light is controlled by drawing up a fold of black tissue over the organ. Squid control the amount of light emitted by squeezing the ink sac between the light organ and the lens. Laboratory studies suggest that the intensity of light and the synthesis of the bioluminescent systems are controlled by oxygen tension. Curiously, synthesis of luciferase, the enzyme involved in the production of light, is enhanced at low oxygen tension in some strains of *Photobacterium*, but in other strains just the reverse is true.

Chapter 9

MICROBIAL ACTIVITIES IN THE BIOSPHERE

Suppose a strange force from outer space suddenly swept up all the microorganisms on earth. Before long, life as we know it would cease. Without microorganisms the essential elements of life would be largely inaccessible. Although on land carbon is photosynthetically fixed by plants, microbial transformation is required before animals can use most of it. In the sea, almost all the carbon is fixed by microorganisms. Utilizable chemical states of nitrogen, phosphorus, and sulfur depend on microbial transformations. The oxygen we breathe was generated over the millenia by microorganisms. When we stop to think about it, microorganisms shaped the biosphere for at least several billion years before higher forms appeared, and since then they have coevolved with these later forms to give us the biosphere in which we now live.

MEASURING MICROBIAL ACTIVITY

Every scale used to examine the natural world reveals its heterogeneity. Minerals are not uniformly distributed, nor are vegetation or animals. At a size scale which is "meaningful" to microorganisms, environmental variation is quite exaggerated, not only in three-dimensional space but also in time. One only has to think of the discontinuous nature of many plant and animal activities (leaf fall, leakage of nutrients from roots, defecation, death, etc.) and the random characteristics of climate to gain an appreciation of this heterogeneity. As an adaptation to nature's variability, many types of microbes distribute their *propagules* (spores, cysts, mycelial fragments, inactive cells, etc.) very widely. As environmental conditions change, microorganisms take appropriate advantage. Because there are often more types of microbial propagules in a community than active microbes, measurements of microbial activity are often difficult.

Overall microbial activity and numbers of microorganisms in soils and water are commonly estimated by measuring either O_2 uptake (or CO_2 production) or total ATP (using the firefly lantern method). In addition to questions about the degree of sensitivity of the techniques and the appropriateness of the constants used in the calculations, the estimates are complicated by the presence in most habitats of large numbers of protozoa and micrometazoa (nematodes, copepods, rotifers, etc.). Particular assemblages of the microbial community are sometimes measured by standard plating of samples on selective media, but inactive propagules skew the estimates. For direct estimations of the numbers and morphological types present, fluorescence microscopy and the scanning electron microscope are appropriate methods.

Although not exempt from problems associated with precision and interpretation, techniques involving radionuclide tracer-tagged compounds (^3H, ^{14}C, ^{35}S, ^{32}P) are valuable for measuring specific microbial activities. One particularly interesting technique uses ^3H-labeled thymidine to monitor DNA synthesis and cell division rates. Cells that have taken up quantities of the label are visualized by *radioautography* in which the radioactivity emitted from the cells strikes photographic film, producing an image of the radioactive cells.

From a variety of methods, estimates of the numbers of microorganisms in various aquatic habitats range from around 1×10^2 to 1×10^9/ml and from around 1×10^3 to 1×10^{10}g in soils.

CARBON CYCLE

All the major mineral cycles (C, N, P, S) involve sequential changes in the oxidation or reduction state of the particular mineral. Photosynthesis is a carbon reduction reaction. Microorganisms are involved in all aspects of the carbon cycle: primary production, transformation, and mineralization. Their importance in each varies with the ecosystem. Figure 9.1 shows the main features of the carbon cycle.

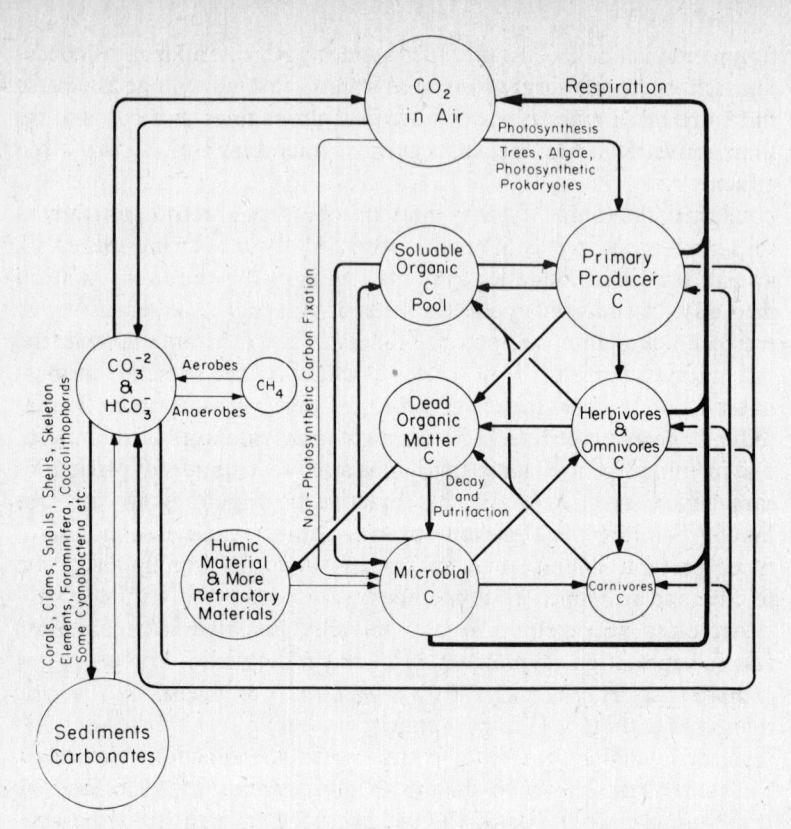

Fig. 9.1. Carbon cycle.

Primary Production. Primary production is an ecological term for the photosynthetic fixation of carbon. It is the first step in the carbon cycle. Slightly over 40% of the total gross annual primary production of carbon of the world ($\sim 1 \times 10^{16}$ kcal/yr) takes place in the upper layers of the world's oceans and large lakes, where almost all of it is microbial. Although a wide range of photosynthetic eucaryotes and procaryotes is found in the oceans, one group of algae, the diatoms, accounts for a high percentage of the total carbon fixed.

Transformation. For almost every organic molecule there is a microorganism capable of transforming or changing it into another.

CELLULOSE DECOMPOSITION. As was mentioned in Chapter 8 in connection with ruminants and termites, cellulose utilization is essentially a microbially mediated process. Since such a high percentage of carbon fixed by the world's plants is synthesized into cellulose and other resistant structural carbohydrates, its degradation is one of the major biosphere processes. Cellulose decomposition takes place in the soil and at the bottoms of lakes and oceans. *Humus* is the organic fraction of the soil which is relatively resistant to decomposition. At the edges of the ocean, particularly in salt marsh or mangrove-bordered estuaries, and in shallow waters that have sea grass meadows, a high percentage (> 90%) of fixed carbon is partially decomposed in the benthos (sea floor). The detrital food web, a major source of secondary production in the sea, is based on the microbial decomposition of the cellulose and related resistant plant structural compounds produced at the margins of the sea. The actual microbial decomposers, bacteria and fungi, are consumed by protozoa and other animals which are, in turn, consumed by still larger ones, and so on.

METHANOGENESIS. Recently, because of the energy crisis, much attention has been given to methanogenesis. Methane production is carried out by a highly specialized group of anaerobic microorganisms (e.g., *Methanobacterium, Methanococcus, Methanosarcina*) which use CO_2 as their terminal electron acceptor.

$$4H_2 + CO_2 \rightarrow CH_4 + 2H_2O$$

In addition the H_2 derived from microbial respiration or fermentation, methanol, formate, acetate, and mercaptan (CH_3SH) can also be converted to methane. It has been estimated that the worldwide annual production of methane by methanogenic bacteria exceeds the annual rate of methane recovery from gas wells.

SYNTHETIC ORGANIC CHEMICALS. In the last few decades, many new synthetic organic chemicals used in textiles, plastics, insecticides, herbicides, and detergents have been produced which are remarkably resistant to microbial attack. Some of them, like the alkylbenzene sulfonates used in household detergents in the 1950s, ran through sewage plants largely unaltered and seeped into water supplies, causing foaming. Newer detergents have linear aliphatic side chains which are more susceptible to microbial attack. Newer federal regulations are forcing manufacturers to pay more attention to biodegradability. A generally useful procedure is the die-away test. The chemical being tested is added to natural soil, water, or sewage and the medium is checked periodically to follow the rate of disappearance.

Mineralization. The skeletons, tests, sheaths, and plates of many aquatic microorganisms are highly mineralized. After they die, or sometimes during their reproduction, their mineralized parts are added to the sediments. Many large deposits of limestone (calcium carbonate, $CaCO_3$) are largely microbial in origin. The Egyptian pyramids were built from blocks of such stone formed from fossilized microbial skeletons. Foraminifera, coccolithophorids, and cyanobacteria are among the more important groups of microbial $CaCO_3$ producers.

Although they are not the only organisms or agents to do so, many types of microorganisms are involved in weathering or dissolution of rocks and minerals. Microorganisms accomplish this usually by the secretion of small quantities of organic acids.

NITROGEN CYCLE

Although nitrogen is the most abundant element in our atmosphere, in its molecular form (N_2) it is not suitable for use by most living organisms. Combined nitrogen—in the form of ammonia, nitrate, and organic nitrogen compounds—is readily assimilated by most organisms, but it is relatively scarce in most natural environments. Figure 9.2 shows the main processes of the nitrogen cycle.

Nitrogen Fixation. Nitrogen fixation is the binding and reduction of dinitrogen (N_2) to a molecular form that makes it readily available for the synthesis of nitrogen-containing organic molecules.

Approximately 85% of the nitrogen fixed from the atmosphere each year is fixed by procaryotic microorganisms. (Symbiotic nitrogen fixation was discussed in the previous section.) A field of clover can fix more than 264 lb of nitrogen per acre per year. Many genera of free-living bacteria (e.g., *Azotobacter, Beijerinckia, Bacillus, Clostridium, Chromobacterium, Enterobacterium, Spirillum, Chlorobium, Chromatium, Rhodopseudomonas*) and heterocysts of cyanobacteria (e.g., *Anabaena, Calothrix, Chloroglea, Nostoc*) also fix nitrogen.

Denitrification. Under anaerobic conditions, nitrate is reduced stepwise to N_2 by some microorganisms (e.g., species of *Bacillus, Moraxella, Xanthomonas, Pseudomonas, Spirillum*). Although denitrification as a process seems wasteful in that it depletes potentially useful nitrogen from the environment, it serves as a regulatory mechanism that prevents the buildup of potentially toxic levels of nitrate.

Assimilatory Nitrate Reduction. Assimilatory nitrate reduction is the process by which nitrate is reduced to ammonia in higher plants, algae, fungi, and many bacteria. The reduction to ammonia makes the nitrogen available for growth through its assimilation into amino acids.

Nitrification. Nitrification is an oxidative process, just the reverse of assimilatory nitrate reduction. Here ammonia is oxidized to nitrate. Animals in the course of their metabolism excrete significant quantities of nitrogenous waste products in the forms of ammonia, urea, or uric acid. Urea and uric acid are rapidly decomposed by groups of microorganisms to CO_2 and ammonia.

The microbial decomposition of animals, plants, and microbes when they die is another major source of ammonia. *Putrefaction*

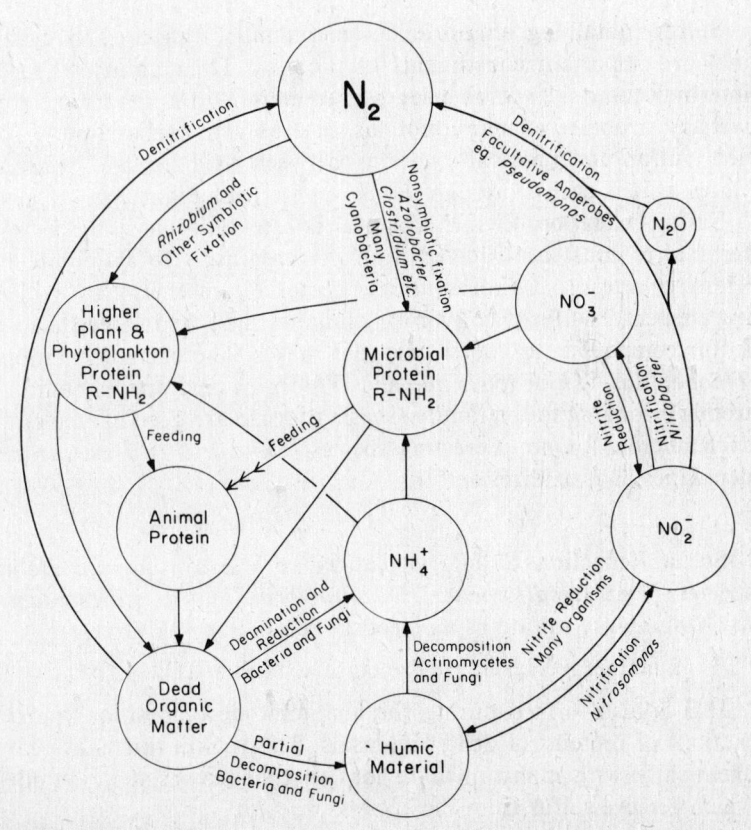

Fig. 9.2. Nitrogen cycle.

(protein decomposition under anaerobic conditions) usually leads to the formation of amines as well as ammonia. Eventually, the amines are oxidized to ammonia by other microbial species.

Nitrification occurs in two steps. First ammonia is oxidized to nitrite (nitrosification) which is then oxidized to nitrate.

$$2NH_3 + 3O_2 \rightarrow 2HNO_2 + 2H_2O \xrightarrow{O_2} 2HNO_3$$

Nitrifying bacteria belong to the family Nitrobacteraceae and are autotrophic. *Nitrosomonas, Nitrosolobus,* and *Nitrosospira* are genera of small oval rods involved in nitrosification. *Nitrobacter, Nitrococcus,* and *Nitrospina* are genera involved in nitrification.

SULFUR CYCLE

Sulfur-containing amino acids (methionine, cystine, and cysteine) are important constituents of proteins. Their ability to form intermolecular sulfhydral bridges is essential to the secondary and tertiary structure of many proteins. Just as with carbon and nitrogen, sulfur undergoes alternations between oxidized and reduced states as it is cycled through the biosphere (See Fig. 9.3).

Sulfur Oxidation. Elemental sulfur is essentially insoluble in water. Sulfur must be oxidized for most organisms to assimilate it. Many colorless and photosynthetic bacteria gain energy by oxidizing elemental sulfur (S), hydrogen sulfide (H_2S), and other reduced sulfur compounds to higher oxidation states. The chemolithotrophs *Thiobacillus, Thiobacterium,* and *Thiospira* and the photolithotrophs belonging to the family Chromatiaceae are good examples of such microbes. One interesting species, *Thiobacillus denitrificans,* denitrifies as it oxidizes sulfur.

$$5S + 6HNO_3 + 2H_2O \rightarrow 5H_2SO_4 + 3N_2$$

Sulfur Reduction. Sulfate is reduced to H_2S by some anaerobic species (e.g., *Desulfomonas, Desulfovibrio,* and *Desulfotomaculum*) which use sulfate as an electron acceptor.

$$2 \text{ lactate} + H_2SO_4 \rightarrow 2 \text{ acetate} + H_2S + CO_2 + H_2O$$

H_2S is also formed during the anaerobic decomposition (putrefaction) of proteins of dead organisms. The H_2S in turn may react chemically with many metallic ions to produce insoluble sulfides which remove sulfur from the ecosystem.

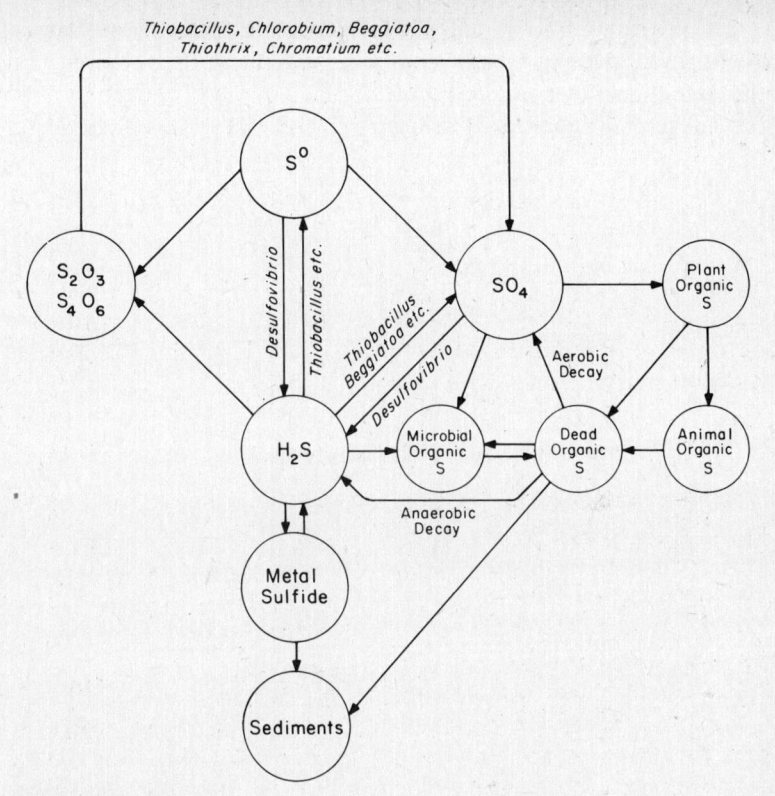

Fig. 9.3. Sulfur cycle.

METALLIC TRANSFORMATIONS AND MICROBIAL LEACHING

Some species of chemolithotrophs (e.g., *T. ferrooxidans*) gain their energy from the oxidation of metallic sulfides. They are usually found in environments with acid pH. The process of solubilizing insoluble minerals like metallic sulfides is called *microbial leaching*.

$$FeS_2 + 3½O_2 + H_2O \rightarrow Fe^{2+} + 2SO_4^{2-} + 2H^+$$

$$CuS + 2O_2 \rightarrow Cu^{2+} + SO_4^{2-}$$

$$Cu_2S + O_2 + 4H^+ \rightarrow 2CuS + 2Cu^{2+} + 2H_2O$$

There are even some patents for processes that use microorganisms for recovering metals from low-grade ores. It is believed that the well-known manganese nodules at the bottom of the sea are also formed through microbial actions.

Chapter 10

APPLIED AND INDUSTRIAL MICROBIOLOGY

Long before people knew of the existence of the invisible life forms we now call microorganisms, they learned how to live in a world filled with them and even to exploit a variety of their activities. Ways were discovered to protect foods from spoilage. The preparation of certain alcoholic beverages predates recorded history. Bread seems to have originated in Egypt about 6000 years ago. Some of our common food preserving and processing is much more recent, although salting, smoking, and drying were known in ancient times. During the Napoleonic wars, the French government offered a prize for a practical method for giving French armies and navies fresh and wholesome food. Voilà! Preserving food in jars was invented, followed a few years later by an English modification, the "tin cannister." Today big industries are built around the manufacture of fermented beverages and foods, antibiotics, chemicals, and other products derived from microbial action.

MICROBIOLOGY OF FOODS

The food industry is concerned with two aspects of microbial activity: food spoilage and food production.

Food Spoilage. Fruits, vegetables, meat, and fish all have a normal microbial flora on their surfaces. While insects, rodents, autooxidants, and enzymatic actions and other processes are responsible in varying degrees for food spoilage, microbial activities on and in foods are significant contributors. Molds belonging to the genera *Aspergillus, Penicillium, Rhizopus, Neurospora,* and *Geotrichum* grow rapidly in bread, foods with high sugar contents, fresh fruits, vegetables, and even cured meat. These organisms cause softening, discoloration, and sourness, and impart the musky odor associated with mold growth. Osmophilic yeasts belonging to the genera *Saccharomyces, Hansenula, Pichia, Candida, Torulopsis,* and *Debaryomyces* grow in jams, honey, fruit juice (sugar 40 to 70%), pickled foods (salt 6 to 16%), and salted meats such as bacon, ham, and frankfurters, causing slimy surface films and imparting odors of esters and aldehydes. Other species of yeasts belonging to the genera *Saccharomyces, Candida Torulopsis, Rhodotorula,* and *Byssolchlamys* are responsible for sourness, bitterness, turbidity, and other undesirable tastes in beer, wine, and canned fruits. Bacteria also do their share of food spoiling. *Clostridium botulinum,* the dreaded producer of botulism toxin, grows well in vacuum-packed and canned meats, fish, fruit, and vegetables, producing putrid and sulfide odors, gas, sourness, and softening. Among others, coliform, lactic acid, and acetic acid bacteria and members of the genera *Bacillus, Micrococcus, Staphylococcus, Pseudomonas, Serratia,* and *Achromobacteria* grow in milk, bread, canned and fresh vegetables, eggs, fruits, meat, and beverages; they produce curdling, sourness, rancidity, discoloration, softening, ropiness, sliminess, blackening, bitterness, and other flavor changes.

Growth of many of these food-spoiling microorganisms can be retarded by storing food in the refrigerator or freezer, smoking, salting, or drying food (e.g., pork, figs, apricots, and raisins), or by lowering the pH in various pickling processes. Some foods have natural substances that retard microbial growth (e.g., onions, garlic, and radishes).

Microbes and the Production of Food Products. Microbes are involved in the production of a number of foods, including milk and milk products, fermented beverages, bread, vinegar, sauerkraut, and pickles.

MILK PRODUCTS. During milking and handling, many microorganisms are introduced into milk. *Streptococcus lactis* and *Lactobacillus* species grow especially well in milk, producing enough lactic acid to lower the pH to 4.8 or below, which coagulates the milk proteins. Pasteurization, the heating of milk to 71.7°C for 15 seconds (HTST method) or 61.7°C for 30 minutes (LTLT method), kills most souring and pathogenic microorganisms. Butter keeps longer than fresh milk because most bacteria do not thrive on high lipid diets. *Pseudomonas putrefaciens* and other pseudomonads commonly make butter rancid by breaking down the glycerides in butterfat. Butyric acid is a common end product in rancid butter.

Fermented Milk Beverages. Buttermilk, yogurt, bus, kefir, koumiss, dadhi, and leben are among the many fermented milk beverages. Buttermilk is prepared by adding a small amount of citric acid (0.15%) and inoculating pasteurized milk with *Streptococcus cremoris* and *Leuconostoc citrovorum*. After incubation, it is beaten until it is smooth and creamy. Yogurt is also made from pasteurized milk in the United States. It is thickened by the addition of rennet or dried milk and soured by a culture containing *Lactobacillus bulgaricus* along with some streptococci and yeasts.

Cheese. The manufacture of cheese is an ancient art that depends not only on the type of milk used but also on the conditions and growth of different species of microorganisms during the curdling and ripening part of the process. Cheeses fall into three broad general categories: (1) natural cheeses; (2) pasteurized process cheeses, cheese foods, and spreads; and (3) whey cheeses. The following steps are involved in the manufacture of cheese.

<u>Setting and Curdling.</u> The milk used in cheese preparation may be obtained from any mammal. Sheep milk, for instance, is used for the manufacture of Roquefort, liptauer, and brinsen. In recent years there has been a tendency to use standard tests to establish milk quality and to pasteurize it. If hard cheeses are being made, some butterfat may be removed. Then *rennet* (rennin) and a starter culture are added and mixed with the milk. Rennin is an enzyme obtained from the stomachs of unweaned calves which curdles milk. If the curd is gently heated ($\sim$ 38°C), strains of *Streptococcus lactis* or *S. cremovoris* are generally in the starter culture. If higher temperatures are used (e.g., 49 to 54°C), starter cultures generally contain mixtures of *Streptococcus thermophilus, Lactobacillus bulgaricus, L. helveticus,* or *L. lactis*. Soft, acid curd, unripened cheeses, such as cottage cheese which are heated to 50°C for 30 minutes have starter cultures containing mixtures of *Leuconostoc citrovorum, L. dextranicum,* and *Streptococcus lactis*. The combined action of acid produced by the bacteria and by rennin curdles the milk protein (casein).

<u>Treating the Curd.</u> The curd is then pressed and sieved so that the whey separates and can be drawn off. The curd is then cut and cooked to develop characteristic texture and body. Cheesecloth or

mechanical presses are used to press the cheese into familiar wheels, bricks, balls, and other forms. When the cheese is hardened, it may be salted or placed in a brine bath. Besides adding flavor, salt controls the growth of undesirable microorganisms.

Ripening. Ripening is due to the slow growth of specific bacteria or molds on the surface and in the interior of cheeses. Some cheeses (e.g., cottage, baker's, cream, and Neufchâtel) are eaten in a fresh, or unripened, state. Hard cheeses such as cheddar, Edam, and Swiss are ripened by microorganisms that cause little proteolysis, whereas soft cheeses like Camembert, Limburger, and Liederkranz are ripened by microorganisms with strong proteolytic and lipolytic enzymes which soften the curd and give it flavor.

The bittersweet flavor of Swiss cheese is due in part to the production of glycerol, propionic, and succinic acids by species of *Propionibacterium* while ripening. The sharp flavor characteristic of Roquefort cheese is developed during ripening by a mold, *Penicillium roqueforti*, which grows within the cheese. Species of the yeast *Geotrichum* form an orange-red slimy coating on the surface of Limburger during initial stages of ripening. Later growth of *Brevibacterium linens* and *B. erythrogenes* changes the slime color to brick red (or red-brown). Camembert is ripened by *Penicillium camemberti* and *Brevibacterium* species. Liederkranz is ripened by *Streptococcus liquifaciens* and *Brevibacterium* species.

FERMENTED BEVERAGES. The origin of fermented beverages is lost in prehistory. Yeasts involved in the production of almost all fermented beverages grow prolifically on the surfaces of most fruits. Most belong to the genera *Saccharomyces, Kloeckera, Candida, Rotulopsis, Hansenula,* and *Brettanomyces. Saccharomyces* species are dominant in warm, dry years while *Kloeckera* species are more common in a cool rainy season. Regional differences in yeast flora are well known.

Wine. In most parts of the world the natural mixed flora growing on the surface of grapes serves as the inoculum for the fermentation of the soluble sugars, glucose and fructose, found in grape juice. The "true wine yeast" *Saccharomyces cerevisiae* var. *ellipsoideus* is the dominant fermenting organism. In regions where the natural yeast flora does not produce quality wine, the *must* (crushed grapes) is treated with sulfur dioxide to kill the natural flora and then inoculated with the desired strain of *S. cerevisiae* var. *ellipsoideus*. Red wines are made by fermentation in the presence of the skins. During aging of sparkling wines, carbon dioxide can be formed by yeasts or a variety of lactic acid bacteria (e.g., *Pediococcus, Leuconostoc,* and *Lactobacillus*). A complex variety of microbial metabolic transformations during the fermentation and aging of wines gives them their distinctive tastes.

Beer. Beers are manufactured from grains like barley, rice, and corn. Unlike fruit, grains contain starch that must first be hydrolyzed (saccharified) to sugars before it can be fermented by *Saccharomyces cerevisiae* or *S. carlsbergensis.* Barley that has been dampened to get it to germinate and then dried for later use is known as *malt.* Germination causes the synthesis of the enzyme amylase which hydrolyzes the starch, yielding maltose and glucose. The malt is ground and suspended in water to allow starch hydrolysis. After saccharification of the starch is completed, the mixture is boiled to denature the enzyme, then filtered. The filtrate is known as *wart.* Hops flowers, which give characteristic bitterness to the brew and which contain some antibacterial substances, are added to the wart, which is then heavily inoculated with good brewery strains of yeast from a previous batch of beer. Strains of brewer's yeast fall into two principal groups: (1) *top yeasts,* those strains which ferment best at relatively high temperatures, 20°C, and produce heavy beers of high alcoholic content (e.g., English ales); and (2) *bottom yeasts,* those strains which are slow fermenters, grow best at lower temperatures, and produce beers of low alcoholic content. The high rate of CO_2 evolution of top yeasts sweeps them up into the top of the vat whereas the bottom yeasts settle to the bottom. In about a week, fermentation is complete. The beer is allowed to settle, filtered to remove the microorganisms and the particulates, then aged, carbonated, and packaged.

Sake. The sugars to ferment sake—oriental rice wine—are released from the grain by the amlyases of a mold *Aspergillus oryzae* which is sown on steamed rice in the first step of manufacture. Yeasts and lactic acid bacteria ferment the sugars in pace with their release from the grain by the mold.

Distilled Liquor. There are four general types of distilled liquors: (1) brandies, from fermented fruit juices; (2) rums, from fermented molasses; (3) neutral spirits, from fermented mashes of mixed grains; and (4) whiskeys, fermented mashes made from only one or mainly one type of grain. Scotch whiskey is made from barley that is allowed to sprout (*malt*). During sprouting, amylases and proteinases hydrolyze the starches and protein in the grain, releasing fermentable sugars. After fermentation with particular "top yeast" strains of *S. cerevisiae,* the brew is distilled and then aged in oak casks for three or more years. Bourbon is a fermented beverage made from a mash of mainly corn with rye, and saccharified with malt. It is also fermented by special strains of *S. cerevisiae.* After fermentation, the brew is distilled and aged in charred oak and hickory barrels.

OTHER FOOD PRODUCTS

Bread. Wheat and other grains contain small amounts of amylases even before they germinate. Small amounts of sugar are released from the starch in moistened flour (dough). When leaven (*Saccharomyces cerevisiae*) is added, the sugar is rapidly fermented. The CO_2 is trapped in the dough and causes it to rise. The alcohol produced during fermentation evaporates during baking.

Vinegar. Vinegar is formed from the ethanol in wine or cider. When vinegar is manufactured in the traditional *Orleans process,* wooden vats are filled with wine, and acetic acid bacteria (*Acetobacter aceti, A. orleanse,* and *A. scheuzenbachi*) grow on the surface in a gelatinous pellicle. Another method for the transformation of alcohol to vinegar uses a *Frings generator*—barrels containing wood shavings which serve as a surface for acetic acid bacteria to grow on. The alcohol substrate is sprayed on the shavings and is oxidized as it trickles through.

Lactic Acid Fermentation. Certain bacteria normally found on vegetables are capable of fermenting the natural sugars in the food, turning them into pleasantly flavored sour foods. To make sauerkraut, cabbage is shredded, salted, and placed into deep fermentation tanks. The salt inhibits the growth of many species of bacteria but not *Leuconostoc, Lactobacillus plantarum,* or *Lactobacillus brevis,* the principal lactic acid producing fermentative organisms used in making sour foods. Lactic acid also protects the food from further microbial attack. Lactic acid fermentations are also important in the manufacture of pickles and Spanish olives.

INDUSTRIAL MICROBIOLOGY

Microbes frequently offer the most economical method for the manufacture of industrial chemicals. In some processes, they offer the only methods. The organisms involved are often mutant strains that are more suitable for the particular industrial process than are wild types isolated directly from nature.

Organic Acids

LACTIC ACID AND ITS DERIVATIVES. Lactic acid and its derivatives are used in the treatment of calcium deficiency and anemia, as a lacquer solvent (*n*-butyl lactate), as a feed supplement, in baking powders, in the acidulation of foods and beverages, in the deliming of hides, in plastics production, and in fabric treatment. Lactic acid can be manufactured from the fermentation of corn and potato starch, molasses, and whey left over from cheese manufacture. The organisms involved in the conversion are *Lactobacillus delbrueckii*, *L. bulgaricus* and other lactic acid bacteria, and the mold *Rhizopus oryzae*.

CITRIC ACID. Citric acid is also used for the acidulation of food and beverages, in pharmaceutical preparations, as a chelating agent, and as a plasticizer. Good yields of citric acid are obtained from substrates of cornstarch, ground corn, beets, black strap molasses, and brewery wastes inoculated with molds such as *Aspergillus niger* and *A. wentii*.

GLUCONIC ACID AND ITS DERIVATES. These are used in pharmaceuticals for chelating iron and calcium, and in the food, feed, leather, photographic, and textile industries. Gluconic acid is produced from glucose-containing substrates during the growth of strains of *Aspergillus niger* or *Penicillium* species.

AMINO ACIDS. Microorganisms are also able to produce large amounts of specific amino acids which can be used to supplement high-protein foods to make them better balanced and increase their nutritional value. The proteins from wheat, rice, and corn are low in lysine. Also, rice protein is low in threonine, corn in tryptophan, and bean and pea in methionine. There are several industrial processes that use microorganisms for the manufacture of lysine. In one process, a mutant strain of *E. coli* that cannot produce lysine is grown aerobically on a substrate of corn-steep liquor, glycerol, and $(NH_4)_2HPO_4$. After incubation, diaminopimelic acid (DAP), an amino acid on the pathway to the synthesis of lysine, accumulates in the medium. The enzyme DAP decarboxylase obtained from *Enterobacter aerogenes* is used to convert DAP to L-lysine. Lysine also is produced directly by mutants of *Micrococcus glutamicus*. Different mutant strains of *M. glutamicus* grown in cultures with a molasses substrate produce commercial quantities of L-glutamic acid.

Strains of *Brevibacterium flavum*, *Arthrobacter* species and *Micrococcus* species also produce commercial quantities of L-glutamic acid in cultures grown in corn-steep liquor and similar media. The concentration of biotin has a significant influence on the yield of glutamic acid. Monosodium glutamate is used as a flavor-enhancing agent and condiment.

ITACONIC AND FUMARIC ACIDS. These acids are used in the manufacture of alkyd resins and wetting agents. They are produced in cultures of strains of the molds *Rhizopus nigricans* and *Aspergillus niger*.

GIBBERELLIC ACID. This is a plant hormone used in the setting of fruit and in seed production. It is obtained from commercial cultures of the mold *Fusarium moniliforme* or *Gibberella fujikuroi*.

Organic Solvents

ACETONE AND BUTANOL. These compounds used as solvents in chemical manufacturing are produced in commercial quantities in cultures of *Clostridium acetobutylicum* and other species.

2,3-BUTANEDIOL. This solvent and chemical intermediate is produced in cultures of strains of *Bacillus polymyxa* and *Enterobacter aerogenes*.

ALCOHOLS. Although most industrial ethyl alcohol is manufactured as a by-product of petroleum cracking, some is produced in commercial fermentations of crude molasses. Small amounts of glycerol, amyl, isoamyl, propyl, butyl, and other alcohols are also recovered from the fermentation vats.

Pharmaceutical Chemicals

DEXTRAN. For use as a blood plasma substitute, dextran is produced in cultures of *Leuconostoc mesenteroides*.

VITAMINS. Vitamin B_{12}, used in the treatment of pernicious anemia and as a feed supplement, is produced in malt extract and corn-steep liquor cultures of strains of *Streptomyces olivaceous*, *Propionibacterium freudenreichii*, and *Bacillus megaterium*. Vitamins B_1 and B_2 are produced in corn-steep liquor cultures of *Ashbya gossypii* and *Eremothecium ashbyi*.

ENZYMES. *Streptokinase*, an enzyme used in dissolving blood clots, and *streptodornase*, an enzyme that degrades DNA to soluble products, can be extracted from cultures of *Streptococcus hemolyticus*. The enzyme *collagenase*, which can be extracted from cultures of *Clostridium histolyticum*, is used to promote healing of wounds and burns. *Takadiastase*, an enzyme that has been used as a digestive aid and bread supplement, can be obtained from cultures of *Aspergillus oryzae*. *Proteases* and *lipases* obtained from cultures of *Aspergillus niger* and *Rhizopus* species are added to some pharmaceuticals as digestive aids.

ANTIBIOTICS. The production of antibiotics is an important part of the pharmaceutical industry. *Penicillin* is obtained from cultures of *Penicillium chrysogenum* grown in sugar-enriched corn-steep liquor. *Streptomycin* is produced in soybean meal extract cultures of *Streptomyces griseus*. The same medium is used to grow *Streptomyces venezuelae* for the production of *chloromycetin*. The various *tetracycline* antibiotics are produced by species of *Streptomyces* grown in a peanut meal extract medium or an enriched corn-steep liquor medium. *Bacitracin* is obtained from cultures of *Bacillus licheniformis* grown in complex organic media.

STEROID TRANSFORMATIONS. Microorganisms are also used for the transformation of one steroid into another. Various steroids have activity as hormones and have been used to stimulate growth of animal food stocks, and in the treatment of various diseases (e.g., shock, arthritis). By adding, withdrawing, or changing the location of attached side groups (e.g., $-OH$, $-COCH_2OH$), the biological activities of the molecules are altered. Different species of molds and *Streptomyces* are the agents for particular transformations. *Corticosterone* ($\triangle 4$-pregnene-11β,21-diol-3,20-dione) is recovered from cultures of *Curvularia lunata*. *Fusarium solani* produces 11-dehydrocortisone.

Insecticides. In recent years there has been an increasing interest in developing microbial insecticides, and much agricultural research is currently directed along this line. One microbe, *Bacillus thuringiensis*, has been successfully employed in the battle against the gypsy moth which has been responsible for extensive destruction of foliage particularly in forests and ornamental trees. *B. thuringiensis* is pathogenic for moth and butterfly caterpillars. A protozoan, *Nosema locustae*, which is pathogenic in grasshoppers, is being used to combat these pests on livestock grazing land.

Industrial Decompositions. Many industrial decomposing processes involve microorganisms in a very general way.

RETTING. Perhaps one of the oldest processes that involve rotting is the controlled decomposition of flax to linen. The stems are soaked in water and the pectinases and other hydrolytic enzymes of butyric acid bacteria soften the plant material and give it its characteristic softness.

SEWAGE TREATMENT. Sewage treatment plants are set up to

maximize the microbial degradation of materials brought to them. Usually there are steps to remove by physical means bulky foreign materials such as bottles, boxes, and paper. Biological oxidations and microbial oxidations take place in large oxidation lagoons, trickling ponds, and various types of open or anaerobic tanks. The microorganisms commonly recovered from sewage plants are members of the families Enterobacteriaceae, Lactobacillaceae, and Pseudomonadaceae and species *Clostridium, Bacteroides, Cytophaga, Micrococcus, Chromobacterium, Aeromonas,* and *Comomonas.* In addition, heavy growth of protozoa, algae, cyanobacteria, molds, and yeasts are found in different parts of various types of plants. Gases such as H_2S, H_2, and CH_4 produced during decomposition can be trapped in plants designed to do so and burned to produce steam for generating electricity to run the various motors in the plant, and for heating the sludge digesters (30 to 40°C) to accelerate microbial activity. One of the final steps in many sewage treatment plants, particularly in the summer, is the treatment of the effluent with chlorine, ozone, or hypochlorite to kill potential pathogenic microorganisms.

METHANE PRODUCTION FROM WASTE. There is increasing interest in building facilities in which methane and other combustible gases are liberated from the decomposition of household garbage and citrus, dairy, poultry, packing house, and cannery wastes. These gases can be used for the generation of steam and electricity.

Microbes and the Production of "Single Cell Protein." There is now some worldwide interest in using as food protein sources certain microorganisms grown in cheap substrates (e.g., oil refinery wastes, methanol, ethanol, corn cobs, wheat straw) in inexpensive and easy to maintain ponds, basins, or fermenters. The rationale is that microorganisms grow 100 to 1000 times faster than higher plants and animals and promise high yield in a short time. The organisms are recovered from the cultures by centrifugation, treated in various ways, dried or cooked, and used as food supplements in animal feeds for cattle, swine, poultry, and fish farming. Among the microorganisms presently being used in actual or pilot plant food manufacture are the alga *Chlorella ellipsoidea,* the cyanobacteria *Scenedesmus acutus* and *Spirulina maxima,* various species of *Pseudomonas* (bacteria), and the yeast *Candida lipolytica.*

Bioassay. Another important use of microorganisms is in the assay of minute quantities ($<$ μg) of biologically active materials.

The test organism is inoculated into a medium that contains all the nutrients required for its growth except the one being assayed. This substance is then added to the medium in certain amounts. The growth of the test organism is proportional to the concentration of the material being assayed. Strains of algae, bacteria, molds, and protozoa have been used as test organisms in various bioassay procedures. Because of their extensive growth requirements, lactic acid bacteria are widely employed for this purpose.

Chapter 11

MICROBIAL PATHOGENS AND DISEASES OF HUMANS

Microbes are always with us—on our skin, in our noses, in our gastrointestinal tracts, in our food. Disease (pathogenesis) occurs when microorganisms become established at sites where they are not normally part of the flora, or the dominant part of the microflora, when they invade body tissues that are normally axenic (germ-free) or when they produce toxic substances that damage their host. Why is it that we can live in harmony with most microbes and yet be harmed by a few troublesome (virulent) ones? Why are some individuals more susceptible to particular microbes than others? How do microbes overcome normal body resistance and produce diseases? Medically oriented microbial research has had a long and distinguished history. In spite of the fact that significant advances toward the answers to these "simple" questions have been made in every decade for the past hundred years, they are still the focus of much active research today.

One of the great microbiologists of the late nineteenth century, Robert Koch, established a set of ground rules for conclusively determining that a particular organism is the causative agent for a disease. These rules, better known as Koch's postulates, are given in Chapter 1. In practice, Koch's postulates cannot always be fulfilled. It is not always possible to grow causative agents in pure (axenic) cultures in artificial media.

FACTORS IMPORTANT IN PATHOGENICITY

Portals of Entry. The body has various kinds of barriers and defenses against infectious agents. Nevertheless, some pathogens do manage to gain entrance. Microbes whose growth is slow or inhibited in certain areas of the body (e.g., *Staphylococcus aureus* on skin) may multiply rapidly in other host environments (e.g., blood and bone).

There are three major portals of entry.

RESPIRATORY TRACT. Inhalation of droplets is the source of infectious diseases such as measles, flu, colds, pneumonia, and tuberculosis.

GASTROINTESTINAL TRACT. Consumption of food contaminated by microorganisms may cause dysentery, cholera, polio, botulism, and other diseases.

SKIN AND MUCOUS MEMBRANES. Direct contact with skin and mucous membranes, wounds, or animal bite inoculation are responsible for tetanus, gangrene, plague, rabies, typhus, malaria, gonorrhea, and syphilis.

Mechanisms of Pathogenicity. Certain properties of pathogens contribute to their invasiveness, or virulence, and enable them to overcome host defensive mechanisms.

CAPSULES. While not every bacterial species that produces a capsule causes disease, many species with capsules (e.g., *Streptococcus pneumoniae, Klebsiella pneumoniae, Haemophilus influenzae, Bacillus anthracis, Yersina pestis*) are pathogenic. Capsules inhibit phagocytosis by the host's leukocytes.

ENZYMES. Many bacteria overcome host resistance and cause disease by the production of extracellular enzymes. Quite a variety are known. *Leukocidin* and *hemolysin,* enzymes produced by staphylococci and streptococci, cause the lysis of leucocytes and red blood cells, respectively. *Hyaluronidase,* an enzyme that digests the intracellular cement, hyaluronic acid, is produced by some pathogenic strains of staphylococci, streptococci, and clostridia. The enzyme *lecithinase,* which hydrolyzes lecithin, an important lipid component of membranes, is secreted by the gangrene-causing *Clostridium perfringens.* The same bacterium also produces *collagenase,* which digests collagen. Coagulase, secreted by *Staphylococcus aureus,* clots fibrin. Other staphylococci and streptococci secrete an enzyme *fibrinolysin* which does the opposite; it lyses fibrin clots. Other enzymes secreted by various pathogenic bacteria depolymerize proteins, nucleic acids, and fats.

TOXINS. Toxic proteins are liberated into the medium by some bacteria during growth. These are *exotoxins* as opposed to *endotoxins,* which are integral parts of the microbial cell. The distinction between invasive enzymes and toxins is based on whether they exert their effects by destroying specific cellular components (*invasive enzymes*) or by interfering with specific cellular functions (*toxins*). Some toxins (e.g., diphtheric, erythrogenic, botulinum toxins) are produced under the direction of bacteriophages (prophages) or plasmid genes.

A very powerful *neurotoxin* which inhibits the release of acetylcholine at myoneural junctions is produced by the anaerobic growth of *Clostridium botulinum* in food. Another powerful neurotoxin which suppresses synaptic inhibition (tetanospasmin) is produced by the anaerobic growth of *Clostridium tetani* in puncture wounds. The diphtheritic toxin produced by *Corynebacterium diphtheriae* inhibits protein synthesis. Gram-negative bacterial walls contain lipopolysaccharide protein complexes (endotoxins) which play important roles in the pathology associated with severe infections. Purified endotoxin injected into an experimental animal causes severe diarrhea, fever, shock, leukopenia, and at high doses, irreversible shock and death. The *enterotoxins* produced by *Shigella dysenteriae* and *Vibrio cholerae* interfere with electrolyte transport.

DISEASES OF HUMANS

Diseases Transmitted by Fecal Contamination of Food and Drink. Pathogens that gain entry by ingestion of fecally polluted food and drink can be debilitating and sometimes fatal. Laws and practices concerning water supplies, sewage, food handling, and swimming water have helped to minimize their transmission.

BACTERIAL

1. *Typhoid fever,* a disease caused by *Salmonella typhi,* is perhaps the best known of these diseases. The bacteria multiply in the intestines and the biliary tract and invade the bloodstream via the lymphatic system. They infect the liver, kidneys, spleen, bone marrow, gallbladder, and occasionally the heart. Symptoms include headache, abdominal pain, weakness, stupor, and fever.

2. Other species of *Salmonella* (e.g., *S. typhimurium, S. schottmulleri, S. choleraesuis*) cause milder *enteric fevers* than those caused by *S. typhi.*

3. *Bacterial dysentery* is caused by *Shigella dysenteriae* which multiply in the gut. The organisms penetrate the epithelial layer and form lesions in the terminal ileum and colon. Symptoms are

cramps, diarrhea, and fever. The bacteria secrete a potent neurotoxin and enterotoxin which can cause paralysis and death when it is injected into experimental animals.

4. *Cholera* is caused by *Vibrio cholerae*. The organisms multiply in the small and large intestines where they are attached to the intestinal mucosa. Symptoms are severe diarrhea in which patients lose as much as 10 to 15 l of fluid per day. Feces known as "rice water stools" contain mucus, epithelial cells, large numbers of vibrios, bicarbonate, and K^+. Death can result from severe dehydration, loss of electrolytes, and electrolyte imbalance. The symptoms are caused by a powerful enterotoxin secreted by the vibrio which binds specifically to a membrane ganglioside and acts to stimulate the activity of the enzyme adenyl cyclase. This, in turn, converts ATP to cyclic AMP (cAMP), an action which stimulates the secretion of Cl^- and inhibits the absorption of Na^+.

VIRAL

1. Human *hepatitis* is caused by at least two different viruses which are clinically distinguished from each other by the length of their incubation period and mode of transmission. *Infectious hepatitis* (hepatitis A) is transmitted from person to person via fecal contamination of food or drink. Incubation averages 35 days. *Serum hepatitis* (hepatitis B) has a longer incubation period (50 to 180 days) and is transmitted by the injection of blood or serum from an infected individual. Both viruses spread throughout the body but most cell damage occurs in the liver. The liver may enlarge, biliary excretion may be blocked, and infected persons may become jaundiced. Asymptomatic individuals can carry the virus in their blood for months or years.

2. *Poliovirus* appears naturally only in humans. Early multiplication occurs in the oropharynx (primarily in the tonsils) and in the intestinal mucosa. The virus spreads to the cervical and mesenteric lymph nodes and then invades the bloodstream. The symptoms in mild cases (aseptic meningitis and abortive poliomyelitis) are fever, headache, drowsiness, sore throat, nausea, vomiting, and stiffness of the neck and back. Paralytic poliomyelitis is characterized by the destruction of large motor neurons in the anterior horn of the spinal cord which give rise to peripheral nerve motor fibers.

3. *Coxsackieviruses* also grow in the pharynx and intestines. Most cases occur in the summer and fall during the swimming season. Two groups of coxsackie viruses are recognized. Group B

causes *aseptic meningitis;* symptoms include fever, headache, nausea, and stiffness in the neck. Group A viruses cause herpangina; symptoms are fever and lesions on the tonsils, soft palate, and pharynx. Patients with herpangina have difficulty in swallowing, lose their appetites, and may have abdominal pain. Summer grippe, devil's grippe (pleurodynia), myocarditis, and pericarditis are also caused by coxsackie viruses.

4. *Echoviruses* cause illnesses similar to coxsackie virus infections. Gastroenteritis resulting in diarrhea can be particularly severe. Infections characterized by skin rashes, fever, and general malaise are common.

PROTOZOAN

1. *Amoebic* dysentery is transmitted by the ingestion of cysts of *Entamoeba histolytica* passed in the feces. Although most of the individuals (90%) that harbor this organism have asymptomatic and undiagnosed infections, severe dysentery and colitis occur in the remaining cases. In typical dysentery, the amoebae penetrate the intestinal mucosa and erode the surface. In mild cases, the symptoms are loose stools containing flecks of blood and mucus. In severe cases, intestinal ulcers are formed which become secondarily infected with bacteria accompanied by diarrhea characterized by the presence of blood and mucus. The ulcers may erode into the adjoining blood vessels permitting the spread of the amoebae to the liver, and occasionally to the diaphragm and lungs.

2. *Amoebic meningoencephalitis* is caused by an amoeboflagellate, *Naegleria fowleri,* which is widespread in moist soils, decaying vegetation, fecal wastes, and contaminated fresh water. Primary infection of the nasal mucosa seems to follow within a week of swimming in contaminated waters. Within 3 to 6 days after the initial symptoms, amoebae get to the brain, causing meningoencephalitis which ends almost invariably in death.

3. *Balantidiasis* is an intestinal disease of hogs and humans caused by *Balantidium coli.* Humans become infected by the ingestion of food and water contaminated with cysts of this large (200-μm) ciliated protozoan. While harmless in most infections, it occasionally causes symptoms similar to acute amoebic dysentery. Occasionally the organism spreads to the lymph nodes, liver, and lungs.

4. *Giardiasis,* an intestinal disease caused by a flagellated protozoan, is also contracted by the ingestion of a cyst. The disease,

which is manifested by diarrhea and abdominal cramps, is particularly severe in humans.

Diseases Caused by Microbial Growth in Food

1. *Staphylococcal food poisoning* is caused by the growth of certain strains of *Staphylococcus aureus* growing in unrefrigerated protein-rich foods such as custard, meat, cream, salad dressing, and deviled eggs. As they grow, the bacteria release a heat-stable exotoxin known as *enterotoxin* which causes vomiting, cramps, diarrhea, and prostration. Complete recovery is generally in 1 or 2 days.

2. *Streptococcal food poisoning* is rarer than staphylococcal poisoning. It is caused by the growth of strains of fecal streptococci, especially *S. faecalis* in unrefrigerated cheeses, evaporated milk, turkey dressing, barbecued beef, turkey à la king, and similar foods. Symptoms are similar to those of staphylococcal poisoning.

3. *Botulism poisoning* is caused by the ingestion of food in which *Clostridium botulinum,* or less commonly, *C. perfringens,* has grown. Of the six different botulism toxins, only three (A, B and E) cause human disease. They are among the most toxic substances known; a milligram of purified toxin A can kill 10 million or more mice. The protein toxin interferes with the transmission of nerve impulses by preventing the release of acetylcholine by various motor nerves. This causes double vision, dizziness, and difficulty in swallowing and breathing.

4. *Mycotoxicosis* is a collective term for diseases induced by the consumption of food that has been made toxic by the growth of various fungi. The best-known disease is caused by *aflatoxin,* a toxin released by *Aspergillus flavus* growing in peanuts, peanut butter, rice, cereal grains, beans, and similar food. Aflatoxin (research indicates that there are 16 or more different aflatoxins) causes severe liver damage resembling alkaloid poisoning. *Ochratoxins* (A and B), produced by *Aspergillus ochraceus* growing in corn, grains, peanuts, Brazil nuts, cottonseed meal, and similar foodstuffs, can also cause liver damage. A mycotoxin released during the growth of *Penicillium toxicarium* on rice causes paralysis, blindness, and death in experimental animals. A mycotoxin produced by *Fusarium graminearum* growing in food is an estrogen that can cause feminization in young males or inflammation of the vulva, vagina, and uterus in females. Fever, bleeding from the nose, throat, and gums, a hemorrhagic rash, and a marked reduction in circulating leukocytes, a disease known as ATA (alimentary toxic aleukia), is caused

by the growth of species of *Fusarium* and *Cladosporium* in grains stored under conditions that favor their growth.

Diseases Transmitted by Exhalation of Droplet Nuclei or Contaminated Dust

BACTERIAL

1. *Diphtheria* is caused by the growth of *Corynebacterium diphtheriae* in the mucous membranes of the upper respiratory tract. A toxin elaborated by the bacilli causes necrosis of the mucosal cells and results in an inflammatory response which leads to a diphtheritic pseudomembrane. The toxin, produced only by phage-infected bacteria, is a single polypeptide chain with enzymatic activity.

2. The severity of *tuberculosis*, a disease caused by *Mycobacterium tuberculosis*, varies with the immunological state of the host, the infecting dose, age, and other factors. Primary infection sites in the alveoli provoke inflammatory responses in the host. If the primary lesion does not heal at once, it develops into a tubercle which consists of a mass of multiplying bacteria, necrotic tissue, and cells associated with the inflammatory process. The tubercle may heal and become infiltrated with fibrous tissue and calcium deposits. The infection can spread to adjacent healthy lung tissue and through the pulmonary vein to many other organs in the body (miliary tuberculosis). Walled-off tubercles are believed to contain viable organisms which persist for the life of the host.

3. *Meningitis* can be caused by a *Haemophilus influenzae* infection in children usually less than 6 years old. The organisms travel from the nasopharynx to the regional lymph nodes where they enter the bloodstream and invade the meninges. The same bacterium can cause acute epiglottitis which can obstruct the passage of air to the point where an untreated child could suffocate.

4. *Whooping cough*, an acute infection of the ciliated epithelial cells of the bronchi and trachea, is caused by *Bordetella pertussus*. Toxins cause local necrosis, inflammation, and fever. Coughing can be so violent that cyanosis ("turning blue"), vomiting, or convulsions follow which completely exhaust the child.

5. *Pneumococcal pneumonia*, an acute lung infection, is caused by the growth of *Streptococcus pneumoniae*. The organism, which has a capsule, resists phagocytosis by leukocyctes. Pneumococci can also invade the sinuses, middle ear, and the meninges. The organisms excrete a hemolysin and a neuraminidase but their role in the

pathogenicity is not known. The inflammatory response of the host leads to the accumulation of alveolar exudates with large numbers of leucocytes (PMNs—polymorphonuclear leucocytes) and red blood cells which make the sputum bloody. Permanent damage to the lungs occurs after recovery because the alveoli become fibrous and inelastic, making breathing difficult.

6. *Anthrax* is a disease of horses, sheep, and cattle. It is spread to humans by the spores that are found in the soil, on vegetation, or animal hides. Spores can be inhaled or can be acquired by contact with hides of infected animals. The bacteria that excyst from the inhaled spores produce hemorrhaging and edema in the respiratory tract. Septicemia can occur which leads to meningitis. Cutaneous infections acquired by contact result in a black necrotic infection. In untreated, severe cases ending in death, the bacteria spread to regional lymph nodes which enlarge and then to the bloodstream. The anthrax bacillus is known to secrete a very powerful exotoxin which interacts with the central nervous system, resulting in acute respiratory failure.

7. *Meningococcal meningitis* (epidemic meningitis, spinal or cerebrospinal meningitis) is caused by *Neisseria meningitidis*. The initial nasopharyngeal infection is usually asymptomatic. Meningococci may enter the bloodstream from the posterior nasopharynx causing *septic meningitis,* which may be explosive, resulting in death in 6 to 8 hours, or may begin with a headache, high fever, and a rash, and cause lesions in the joints, lungs, and adrenals. Meningococci may cross the blood-brain barrier and infect the meninges, causing severe headache, stiff neck, vomiting, and delirium. The hemorrhagic lesions in the skin and internal organs are believed to be due to the release of endotoxin.

8. *Psittacosis-ornithosis* is an infection acquired by the inhalation of dried bird feces containing *Chlamydia psittaci*. While many cases are asymptomatic, a fatal pneumonia may develop.

9. *Q fever* is an unusual rickettsial disease caused by *Coxiella burnetii*. Unlike other rickettsiae which are transmitted by arthropods, Q fever is acquired by inhalation of dried tick feces. Pneumonia with a mild cough is accompanied by chills, headaches, malaise, and weakness.

VIRAL

1. *Influenza* is caused by three types (A, B, C) of orthomyxoviruses with envelopes containing spikes of neuraminidase and hem-

agglutinin. The hemagglutinin allows the virus to attach to the surface of erythrocytes and the neuraminidase aids in the release of the virus. Symptoms are fever, chills, headache, general muscular aching, and loss of appetite. Although the virus usually infects the epithelial surfaces of the upper and middle respiratory tract, in severe cases the virus infects the lung epithelial tissue causing viral pneumonia. Reduced resistance to shifts in antigenic types in the viral population permit worldwide epidemics (pandemics) to take place every 2 to 5 years.

2. *Parainfluenza viruses,* which belong to a different family of viruses, the Paramyxoviridae, also produce upper respiratory infections. In infants and children they may cause pneumonia. Types 1 and 2 of the virus frequently invade the larynx of infants, causing the croup.

3. *Mumps* is also caused by a paramyxovirus. It multiplies in the upper respiratory tract and in the local lymph nodes from which it enters the bloodstream. Most frequently it causes painful swelling of the parotid glands but it may also develop in the meninges, pancreas, testes, ovaries, or the heart.

4. *Measles* (rubeola) is another disease caused by a paramyxovirus. With a sore throat as an early symptom, infections begin in the upper respiratory tract and conjunctiva. After the virus enters the bloodstream, high fever, coughing, delerium, conjunctivitis, a rash, and photophobia (sensitivity to light) become clinical symptoms. Severe cases can have pneumonia, ear infections, or encephalitis.

5. *Rubella* (German measles) is a rather mild respiratory disease caused by a togavirus. After replication in the cervical lymph nodes, the virus is disseminated throughout the bloodstream, causing mild fever and rash. The virus can cross the placental wall, invade the fetus, and cause congenital defects.

6. *Common colds* are caused by a large number of small RNA viruses belonging to the picornavirus family. They are collectively known as *rhinoviruses.* Infections are restricted to the cells of the upper respiratory tract.

FUNGAL

1. *Blastomycosis* is caused by *Blastomyces dermatitidis,* an organism that survives in the soil for several months after being deposited there by animal feces. The spores are inhaled, causing a pulmonary infection not unlike tuberculosis. Untreated infections

can progress to acute lobar pneumonia and spread via the bloodstream to bone, skin, and rarely, internal organs.

2. *Cryptococcosis,* caused by *Cryptococcus neoformans,* is acquired by the inhalation of dried infected pigeon feces. Primary infections are in the lungs. Pneumonia may be followed by the spread of the organisms through the bloodstream to the brain and meninges.

3. *Histoplasma capsulatum* is also acquired by inhalation of spores that are associated with bird and bat feces. *Pulmonary histoplasmosis* resembles tuberculosis. In some cases, the organisms spread to the liver, spleen, and lymph nodes.

Diseases Transmitted by Direct Contact

BACTERIAL

1. *Brucellosis* (Malta fever, Cyprus fever, undulant fever) is caused by aerobic bacteria belonging to the genus *Brucella.* Most human infections of brucellosis are acquired through the skin or conjunctiva by direct contact with infected animals (goats, pigs, cattle, and sheep). Persons can also contract the disease by ingestion of contaminated dairy products. Primary invasion spreads via the lymphatic system. Even though the bacteria are phagocytosed by macrophages, they are able to multiply within these cells. The bloodstream is invaded. As macrophages die, brucellae are released to form lesions in the liver, spleen, bone marrow, and kidney. Symptoms in humans are malaise, weakness, and undulating diurnal fever.

2. *Gonorrhea,* a disease caused by *Neisseria gonorrhoeae,* is, with few exceptions, acquired by sexual contact with an infected individual. Primary local infections occur in the mucous membranes of the genital tract. Men may have asymptomatic infections but they usually have acute urethritis with painful urination and a purulent discharge from the penis. Gonococci may also invade the prostate gland and the epididymis. The disease is more likely to be asymptomatic in women. In severe infections, gonococci infect the vagina, cervix, and fallopian tubes which may result in sterility. Bacteremia can lead to lesions in the skin, heart, eyes, and joints. Infants can acquire serious eye infections during their birth from diseased mothers.

3. *Syphilis,* a disease caused by the spirochete *Treponema pallidum,* is also contracted by sexual contact. Invasion of the mucous membranes produces localized ulcerations known as *chancres.* The

organisms migrate from the primary lesions to the regional lymph nodes and then are transported throughout the body by the bloodstream. A skin rash and lesions on the mucous membranes, eyes, bones, and central nervous system are symptoms of secondary syphilis. Tertiary syphilis occurs years (5 to 40) after the initial infection. Lesions occur in the cardiovascular system or the central nervous system, causing aortic aneurysms or paresis. *T. pallidum* can cross the placenta and either kill the fetus or give it congenital syphilis.

4. *Yaws* is a tropical disease caused by a spirochete, *Treponema pertenue,* closely related to the one that causes syphilis. It is spread from one person to another either by direct contact with skin ulcers or by flies. Secondary and tertiary lesions of the skin and bones, particularly of the face, can occur months or years after the initial infection.

5. *Trachoma, inclusion conjunctivitis,* and *lymphogranuloma venereum* are diseases caused by different strains of the tiny bacterium *Chlamydia trachomatis.* Trachoma, a conjunctival disease, is transmitted by direct contact with fingers or contaminated towels and clothing. Although the disease begins with an inflammation of the conjunctiva, there is an accumulation of lymphocytes and macrophages to form red or yellow follicles on the conjunctiva. Vascularization and infiltration of the cornea take place, which can lead to partial or complete blindness. Inclusion conjunctivitis is both a venereal disease and a disease of the conjunctiva. The genital symptoms are usually very mild. However, infants acquire conjunctivitis at birth from diseased mothers. Adults can also get eye infections from swimming in waters contaminated by genital or eye secretions. Inclusion conjunctivitis is less severe than trachoma and does not lead to blindness. Lymphogranuloma venereum is an uncommon venereal disease. Primary lesions are small papules which ulcerate and then heal. The organisms migrate to the inguinal or iliac lymph nodes which enlarge (buboes) and become tender. Lymph channels can be completely obstructed, causing enlargement of the genitalia.

6. Species of *staphylococcus* are ubiquitous on human beings. Infections by virulent strains of *S. aureus* are acquired from lesions, fomites (towels, sinks, door knobs) contaminated from them, the skin, respiratory tract, and aerosols. They cause a number of different infections ranging from localized furuncles (boils), car-

buncles, osteomyelitis, pneumonia, impetigo (also caused by other bacteria), and several other less common infections. As mentioned earlier in this chapter, staphylococci elaborate a number of toxins and enzymes which aid them in their pathogenesis. These are:

a. *Coagulase,* an enzyme-like substance that clots plasma. The action of coagulase may be to coat the organisms with fibrin to inhibit their phagocytosis.
b. *Staphylococcal hemolysins* (alpha, beta, gamma, and delta) lyse red blood cells and skin (particularly alpha and delta) cells.
c. *Leukocidin* is a factor that kills white blood cells of many mammals.
d. *Exfoliatin* is an exotoxin, genetically coded by a plasmid, which causes acute exfoliative dermatitis. Known as the scalded skin syndrome, the skin becomes red and sloughs off.
e. Various strains of *Staphylococcus aureus* may also secrete enzymes such as *Penicillinase* (destroys penicillin), *hyaluronidase* (a spreading factor), *lipases,* and *staphylokinase* (dissolves fibrin clots).

7. *Streptococci* (e.g., *Streptococcus pyogenes, S. faecalis, S. salivarius*) are widely distributed in nature, are members of normal human flora, and are associated with some important human diseases. More than 90% of streptococcal diseases are caused by group A beta-hemolytic (see below) strains of *S. pyogenes.* Streptococcal diseases are divisible into broad categories: (1) primary infections (suppurative) and (2) late nonsuppurative complications. The portal of entry of the disease determines the clinical picture. The most common streptococcal infections are *nasopharyngitis* (sore throat) and *tonsillitis.* If the patient has no antitoxic immunity and if the strain produces erythrogenic toxin (see below, *scarlet fever*), rash occurs. Local infection of the skin leads to the development of *impetigo,* a superficial rash of small blisters that become crusty. *Erysipelas* is an infection of the skin or mucous membrane characterized by a spreading inflammation. *Puerperal fever* (childbirth fever) is an infection of the womb (endometritis) after delivery. Acute *glomerulonephritis,* an inflammation of the glomeruli of the kidney, develops in some persons 3 weeks following streptococcal infections with certain types (2, 4, 12 and 49) of streptococci. *Rheumatic fever,* a disease that results in damage to heart muscles

and valves, follows a few weeks (1 to 4 weeks) after untreated pharyngitis by certain strains of streptococci. *Erythema nodosum* is a skin condition in which small red nodules appear under the surface of the skin following streptococcal infections, tuberculosis, or coccidioidomycosis. As mentioned earlier in this chapter, streptococci elaborate a number of substances which aid them in their pathogenesis. These are:

  a. The *erythrogenic toxin* causes the rash of scarlet fever. It is produced by strains harboring a lysogenic phage.
  b. *Streptokinase* (fibrinolysin) acts by catalyzing the conversion of an inactive plasma component, plasminogen, into plasmin, an active proteolytic enzyme that digests fibrin.
  c. *Streptodornase* is an enzyme that depolymerizes DNA and is believed to contribute to the spreading of the infections by liquefying DNA from the host's leukocytes in pus.
  d. *Hyaluronidase* deploymerizes hyaluronic acid and aids spreading of streptococci.
  e. Several *hemolysins* that hemolyze red blood cells are released by streptococci. Incomplete lysis of erythrocytes with the formation of green pigment is called alpha (α) hemolysis. Complete disruption is known as beta (β) hemolysis. Beta hemolytic streptococci elaborate two hemolysins, streptolysin O and streptolysin S.
  f. Some strains of streptococci also elaborate *diphopyridine nuclease, proteinases,* and *amylases.*

VIRAL

1. *Herpes* type 1 virus is spread by kissing (saliva) and by improperly washed utensils or dishes and similar objects that might be contaminated by saliva. Type 2 herpes virus is spread venereally or during birth. Type 1 virus is the agent that causes *cold sores* (herpes labialis, herpes febralis); acute *herpetic gingivostomatitis,* a disease causing ulcerative lesions on the mucous membranes of the mouth; *eczema herpeticum,* a disease characterized by extensive vesiculation of the skin over much of the body; *keratoconjunctivitis,* an infection of the eye which causes an extremely painful ulceration of the cornea, which can lead to permanent opacification and blindness; and *herpetic encephalitis* (meningoencephalitis), a disease in which the virus causes lesions in localized areas of the brain, which may result in paralysis, mental retardation, or death. Type 2 herpes

virus is the agent that causes *herpes progenitalis* (genital herpes), a disease usually resulting in mildly discomforting lesions on the vagina, cervix, or vulva and painful lesions on the glans penis, and *neonatal herpes,* a disease of the newborn which results in loss of weight, vomiting, diarrhea and, in severe cases, brain damage.

2. *Varicella* (chicken pox) is a mild, highly infectious disease principally of children, characterized by skin and mucous membrane eruptions. In *zoster* (shingles) there is an inflammation of dorsal nerve roots and sensory ganglia as well as skin lesions. The varicella-zoster virus is closely related to the herpes simplex virus, as are the Epstein-Barr (EB) and infectious mononucleosis viruses.

3. *Warts* are spread by direct or indirect contact. Though the virus has not been grown in tissue culture or in laboratory animals, filtrates recovered from warts have produced warts in volunteers.

4. *Smallpox* (variola) begins as an infection of the mucous membranes of the upper respiratory tract and regional lymph nodes where it multiplies before entering the bloodstream. The virus localizes and replicates in the reticuloendothelial system before invading the skin where it produces the typical smallpox lesions. Fever, headache, and loss of appetite are other symptoms. The disease is infectious after the appearance of clinical symptoms. Smallpox is transferred directly from person to person or through contaminated fomites.

FUNGAL

1. *Ringworm* of the scalp (*tinea capitis*) is caused by species of *Microsporum* and *Trichophyton.* The fungi invade the keratinized superficial layers of the scalp and hair, causing scalp inflammation and itching.

2. Species of *Microsporum* and *Trichophyton* also cause ringworm of the body (*tinea corporis*), a dermatophytosis involving nonhairy areas. The lesions are reddened, scaly, papular eruptions.

3. *Athlete's foot* (ringworm of the feet, *tinea pedis*) is caused by a number of species of *Microsporum, Trichophyton,* and *Epidermophyton.* Infected individuals have itching and burning between the toes, and peeling, cracked skin.

4. *Oral thrush* and *vaginal candidiasis* are diseases caused by *Candida albicans,* a common, usually commensal yeast found in the mouth, intestinal tract, and vagina. The oral disease is characterized by inflammation of the oral surfaces (gingiva, palate, tongue, etc). Newborn infants can become infected from their mothers at

birth. Vaginal candidiasis is characterized by inflammation and by an odoriferous white discharge. *Candida albicans* infections can also occur on the skin, particularly on the feet, hands, groin, axillae and, rarely, systemic following trauma or surgical procedures which are treated with antibiotics, steroids, or other immunosuppressive drugs.

5. *Aspergillosis* is a localized infection of the ear caused by the fungus *Aspergillus niger,* growing on the wax of the ear. The same organism can cause pulmonary infection as a secondary invader or in patients treated with immunosuppressive drugs.

PROTOZOAN. *Trichomonas vaginalis* causes trichomoniasis, an annoying venereal disease. In males the disease is usually symptomless, although sometimes urethritis, epididymitis, prostatovesiculitis, and dysuria (painful urination) occur. Most women experience only a minor vaginal discharge. In heavier infections, there is a vaginitis with a copious, foul-smelling yellowish-greenish discharge accompanied by almost intolerable intense pruritus (itching) and burning.

Wound Infections. Many of the organisms discussed in previous categories (e.g., *Staphylococcus, Streptococcus, Candida*) can also cause infections if introduced into wounds.

BACTERIAL

1. *Tetanus* is caused by the introduction of spores of the anaerobic bacterium *Clostridium tetani* into puncture wounds, burns or animal bites. The organism itself is widely distributed in the soil, and in the feces of farm animals. Germination of the spores is aided by dead tissue and associated pyogenic infections at the site of the wound. A powerful neurotoxin, *tetanospasmin,* is believed to travel from the wound along nerve fibers to the spinal cord anterior horn cells where it causes convulsive contraction of the voluntary muscles. The purified toxin is a heat-labile protein, a single milligram of which can kill more than 2×10^7 mice.

2. *Gas gangrene* is caused by a number of species of *Clostridium*, (*C. perfringens, C. novyi, C. septicum, C. histolyticum,* and others). The spores of these bacteria are common in the soil and herbivore feces. Deep wounds and dead tissue favor germination. As clostridia multiply they ferment carbohydrates present in tissues and produce gas that causes painful pressure in the infection. As mentioned earlier in this chapter, gangrene associated bacteria produce toxins and extracellular products that aid in their pathogenesis. The alpha toxin of *C. perfringens* is a lecithinase that attacks

cell membranes. They also produce hyaluronidase, collagenase, proteinase, and deoxyribonuclease.

3. *Leptospira icterhemorrhagiae* cause chronic kidney infections in many animals including rodents, dogs, pigs, cattle, and wild animals. Humans become infected through skin contact with urine from infected animals or by exposure to urine-contaminated water or soil. Ingestion of bacteria in contaminated water and food is also a route of infection. The organism multiplies in the blood and infects the liver, kidney, lung, and meninges. Symptoms include fever, headache, muscular pain, jaundice, and hemorrhage in the skin and subcutaneous tissues.

FUNGAL. *Sporotrichosis,* caused by *Sporothrix schenckii,* is a skin disease of farmers, gardeners, greenhouse workers, and others who are constantly exposed to soils. The organisms are introduced into the subcutaneous tissues by wounds from splinters, thorn pricks, and similar accidents. After gaining entry, they multiply and a subcutaneous nodule forms, which develops into a necrotic ulcer. As initial lesions heal, new nodules and ulcers develop along lymph channels.

Diseases Caused by Animal Bites
BACTERIAL

1. *Bubonic plague,* a potentially fatal disease caused by *Yersinia pestis* (formerly called *Pasteurella pestis*), is spread by the bites of infected rat fleas. The organisms migrate from the site of the flea bite to regional lymph nodes, causing their enlargement. After leaving the regional lymph nodes, *Y. pestis* spreads via the bloodstream to the spleen, liver, lungs, and other organs. Subcutaneous hemorrhages give the disease its popular name "black death." The lungs become involved and the disease can be spread via respiratory discharges.

2. *Tularemia,* a disease caused by *Francisella tularensis,* is spread by the bites of ticks and flies associated with wild animals such as rabbits, muskrats, and other game. The course of the disease follows the route of entry of the pathogens. The most common form of the disease, *Ulceroglandular tularemia,* begins as ulcerated sores of the fingers and hand, usually acquired by tick bites during the skinning and dressing of a game animal. The organisms spread to the regional lymph nodes and may travel through the bloodstream to the liver and spleen; occasionally they enter through the eye causing a *conjunctival* ulcer (ulcero-ocular tularemia). The dis-

ease is sometimes acquired by the ingestion of contaminated food and drinking water, causing symptoms similar to typhoid fever (*typhoidal tularemia*).

3. *Epidemic typhus,* caused by *Rickettsia prowazekii,* is transmitted to humans by body lice (*Pediculus humanus*). The rickettsiae, which live in the alimentary tract of the louse, may be introduced during a bite, or when the feces of a louse or a crushed louse is scratched into the skin. After an incubation period of 6 to 15 days, fever, and a rash on the trunk and extremities develop. The rickettsiae have an affinity for the endothelial cells of the small blood vessels; these cells may become detached and enter the bloodstream, causing obstructions that can lead, particularly in cold weather, to gangrene of the feet and fingers. Myocardial or neurological involvement can be fatal in older patients.

4. *Endemic typhus,* which is a mild form of typhus, is caused by *Rickettsia mooseri.* It is transmitted to humans by the rat flea. DDT has been effective in controlling both diseases during unsanitary conditions such as those experienced during wars and natural disasters.

5. *Scrub typhus,* a disease that had particular importance for the United States during World War II and the Viet Nam conflict, is caused by *Rickettsia tsutsugamushi.* The disease and the rodent mites that spread it are found in China, eastern Asia, and South Pacific islands. Symptoms include severe headache, chills, rash, fever, stupor, and prostration. A primary lesion, called an eschar, occurs at the site of the bite. The spleen often becomes enlarged.

6. The body louse is also the vector for another rickettsial disease, *trench fever.* The causative agent is *Rickettsia quintana.* It is a relatively mild disease which was fairly common in the two world wars.

7. *Rocky Mountain spotted fever (RMSF), Asian tick typhus, African tick typhus, Rickettsial pox,* and several less common fevers are rickettsial diseases of wild animals occasionally transmitted to humans by ticks and mites. RMSF is harbored in the United States in rabbits, raccoons, woodchucks, deer, and small rodents. Two to five days after the tick bite, a rash, fever, headache, and lymphadenopathy develop.

8. *Relapsing fever* is caused by any one of 11 different species of *Borrelia* (e.g., *Borrelia recurrentis*). They are spread by the body louse or by ticks. The bacteria enter the bloodstream and cause

lesions in the spleen, liver, kidneys, and gastrointestinal tract. The symptoms are fevers of 4 to 5 days, an afebrile period of a week to 10 days, then a relapse to a febrile period. There may be as many as 3 to 10 relapses before complete recovery.

VIRAL

1. The *rabies* virus seems capable of infecting most mammals. Reservoir animals vary in different parts of the world but in North America include dogs, cats, bats, skunks, foxes, coyotes, and raccoons. The infection is transmitted by the bite of a rabid animal. It has been shown experimentally that animals caged in caves inhabited by rabid bats can contract the disease via the respiratory tract. After the virus is introduced into the bite wound, it progresses along nerve paths to the central nervous system where it produces an often fatal encephalitis. Initial symptoms of headache, nausea, excitation, nervousness, and sore throat are followed by difficulty in swallowing ("hydrophobia"), convulsions, coma, and death. The incubation period can be extremely long (2 to 16 weeks or longer).

2. *Yellow fever, dengue, West Nile fever, equine encephalitis,* and a number of other types of encephalitis are caused by arboviruses (short for *A*rthropod *bo*rne viruses), a collective term for single-stranded RNA-containing viruses belonging to the families Togaviridae and Bunyaviridae. Birds and mammals are the natural reservoirs. Humans are accidental hosts when they enter an area where a usually wild animal and blood-sucking arthropod cycle is maintained. Mosquitoes are the usual vector. The symptoms of yellow fever are fever, prostration, hepatitis, and nephritis. Headache, fever, myalgia, prostration, and rash are the symptoms of dengue. Eastern equine encephalitis produces lesions throughout the brain, causing a high mortality rate and neurological impairment of the survivors.

PROTOZOAN

1. *Malaria* is caused by species of *Plasmodium (P. vivax, P. ovale, P. malariae* and *P. falciparum)* which parasitize red blood cells. Duration and intensity of symptoms vary with each particular species. Symptoms are generally chills and fever, at more or less regular intervals, followed by sweating. Bites by species of the *Anopheles* mosquito transmit the disease. The life cycles of malarial parasites are quite complex. After a human is bitten, the first stage (pre-erythrocytic) of development takes place in the parenchymal cells of the liver. The *sporozoites* asexually reproduce numerous

```
Mosquito                              Human
(sexual                              (asexual
 cycle)                               cycle)


Sporozoites    Mosquito bite    Sporozoites
in salivary
  gland                               │
                                      ▼
                              Pre-erythrocytic
                              stages (in liver
                              parenchymal cells)
                                      │
                                      ▼
Sporozoites                     Merozoites
                              (in red blood cells)
    ▲                                 │
                                      ▼
 (Multiple                      Ring trophozoites ──┐
  division)                            │            │
                                       ▼            │
Oocyst (outer layer            Mature trophozoites  │
of stomach wall)                       │            │
    ▲                                  ▼            │
Ookinete in intestine              Schizont         │
                                                    │
  Microgamete  ◄──────────── Macrogamete  Macrogamete
Zygote           Mosquito bite
  Macrogamete ◄──
```

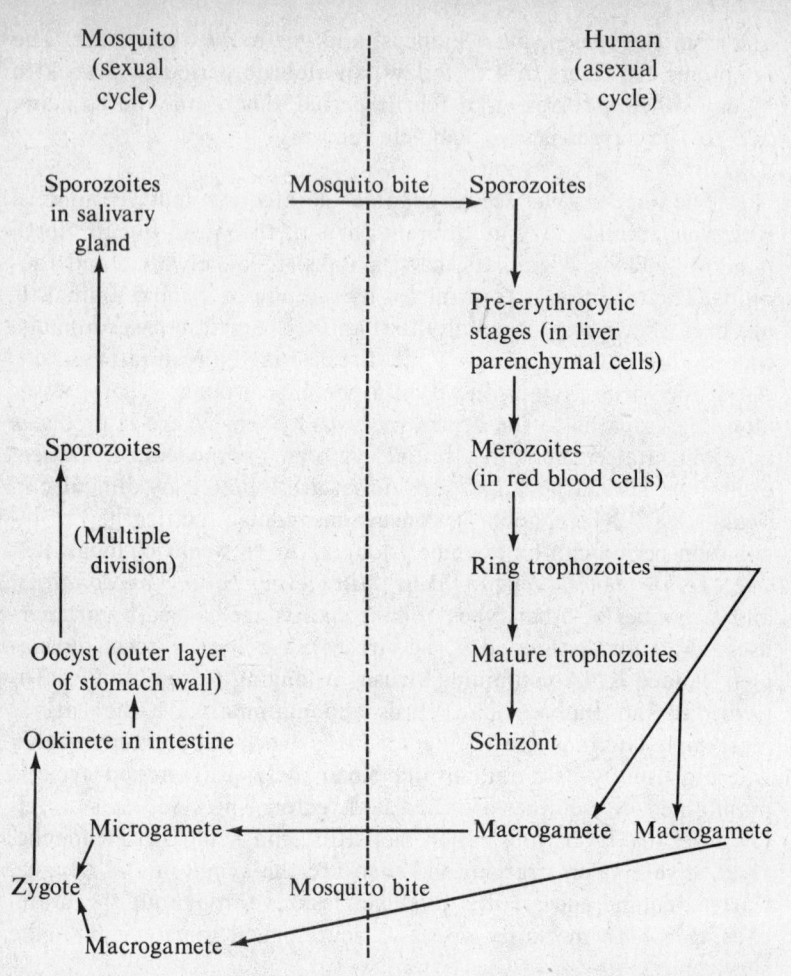

Fig. 11.1. Life cycle of a malarial parasite.

progeny (merozoites) which leave ruptured liver cells, enter the bloodstream, and invade erythrocytes. The merozoites become *trophozoites* which multiply and rupture their host erythrocytes, releasing new broods of merozoites synchronously in a process called the *erythrocytic cycle*. During the erythrocytic cycles, some merozoites that enter red cells become differentiated as either *male* or *female gametocytes*. The next stage in the life cycle takes place in the mosquito. When the gametocytes are ingested, they become male and female gametes which fuse in the stomach of the mosquito to form zygotes. The zygotes undergo morphogenesis to become a new stage, the *ookinete*. These round up outside the stomach to become *oocysts*. Hundreds of spindle-shaped sporozoites develop within each oocyst. Sporozoites migrate throughout the mosquito's body and many enter the salivary gland ready to renew the cycle again.

2. *West African sleeping sickness* is caused by the flagellated protozoan *Trypanosoma gambiense*. A more virulent form of this disease, *East African sleeping sickness,* is caused by a related organism, *Trypanosoma rhodesiense*. Both organisms are transmitted by the bites of infected tsetse flies. An ulcer may develop at the site of the bite but the fever symptoms do not appear until 2 to 3 weeks later, when the organisms have invaded the bloodstream and lymph nodes. After several days to a week, the fever subsides. Several weeks may elapse before the attack returns. As the disease progresses, the face may swell, the heart may be damaged, and the trypanosomes invade the central nervous system, causing meningoencephalitis that leads to mental deterioration, slurred speech, convulsion, paralysis, coma, and eventually death.

3. *Chagas' disease* (American trypanosomiasis) is caused by a related species, *T. cruzi*. A wide variety of wild animals (e.g., rodents, armadillos, opossums) serve as reservoir species. Humans acquire the disease from infected reduviid (triatomid) bugs. The flagellate grows in the gut of the bug and is passed out in its feces. When the bugs bite humans, some fecal contamination of the wound occurs, admitting the trypanosome into the body. From the site of the bite the organisms travel to the regional lymph nodes and spread from there to other reticuloendothelial organs. The heart, liver, and macrophages in the spleen are commonly infected. Growth of the flagellates in the heart muscle produces an inflammatory response and an enlarged heart. The flagellates may invade the central nervous system of infants and young children, producing a fatal meningoencephalitis.

4. *Cutaneous leishmaniasis* (oriental sore) is caused by another flagellate, *Leishmania tropica*. It is transmitted by the bite of a sand fly (*Phlebotomus*). The infection is localized to the site of the bite. An ulcerated lesion develops which heals after many months, leaving disfiguring scars.

5. *Mucocutaneous leishmaniasis,* caused by *L. braziliensis,* is initially similar to the cutaneous disease. After the healing of the

lesions, the organisms reappear, causing new lesions in the mucous membranes in the nasopharynx. Secondary bacterial infections can be fatal.

6. Sand flies also transmit *Kala azar (visceral leishmaniasis),* an often fatal Near and Far Eastern disease caused by another species, *L. donovani.* The flagellates spread from the site of infection to multiply in reticuloendothelial cells, especially macrophages in the spleen, liver, lymph nodes, and bone marrow. The spleen and liver usually become enlarged. Untreated patients become progressively emaciated, rendering them subject to secondary infections.

HOST DEFENSE MECHANISMS

A somewhat arbitrary distinction is made between the "natural," or constitutive, mechanisms by which a host resists most broad groups of pathogenic microorganisms and the inducible mechanisms for acquired resistance to a particular pathogen or closely related groups of pathogens. The former mechanisms are commonly referred to as *resistance,* the latter, *immunity.*

Natural Resistance. Natural resistance includes both internal and external mechanisms of defense.

EXTERNAL DEFENSE MECHANISMS (FIRST LINE OF DEFENSE)

Mechanical Barrriers. The skin and mucous membranes of the body are barriers that prevent the entrance of most microbial species. Sweat and the secretions of the sebaceous glands, which contain NaCl, lactic acid, and fatty acids, inhibit the growth of many types of microorganisms. The mucus secreted by the membranes that line the respiratory, digestive, and urogenital tracts traps microorganisms. Tears, saliva, nasal secretions, and mucus contain an enzyme, lysozyme, which can dissolve the peptidoglycan in the walls of bacteria by hydrolyzing the linkages between the N-acetylmuramic acid and N-acetylglucosamine. This causes the cell wall to break into fragments which, under usual osmotic conditions, results in the rupture of the cell.

Mechanical Action. Peristalsis, coughing, sneezing, perspiring, and tears provide mechanical flushing of microorganisms to places where they can be expelled or destroyed.

INTERNAL DEFENSE MECHANISMS (SECOND LINE OF DEFENSE). Once foreign agents penetrate mechanical barriers, various elements of the reticuloendothelial, vascular, and lymphatic systems (and many tissues as well) respond relatively nonspecifically. Inflammation is a tissue reaction triggered by foreign materials. Leucocytes migrate into the affected area and increased blood flow produces swelling, reddening, heat, pain, and tenderness.

Blood Components (Phagocytes). Blood contains three main types of cellular elements: red blood cells (erythrocytes), white blood cells (leucocytes), and platelets. White blood cells that can engulf and destroy foreign cells and material are called phagocytes.

Types of Phagocytes. Two types of leucocytes are active phagocytes: the neutrophils and the macrophages.

Neutrophils (PMNs). The neutrophils, also known as microphages, polymorphonuclear cells, PMNs, PMLs, and polys, are the major circulating leucocytes (65 to 70%). PMNs contain membrane-bound organelles, the *lysosomes,* that when stained, give the cell a granular appearance. Lysosomes contain enzymes and antimicrobial substances for the killing and degradation of foreign cells. During an inflammatory reaction, *cytotaxins* attract or stimulate migration of PMNs toward the inflamed site. Cytotaxins include (1) some components of the complement system (see below), (2) bacterial products, (3) products from damaged tissue cells, and (4) cyclic AMP and other substances.

Macrophages (Monocytes). Macrophages are large, nongranular-appearing leucocytes. They may be wandering or fixed. The system of fixed macrophages is often referred to as the *reticuloendothelial system.*

Phagocytosis. Phagocytosis is the ingestion and digestion of foreign cells or particles by PMNs and macrophages. It is a two-step process: (1) initial contact, recognition, and binding (attachment) and (2) ingestion, with the formation of a phagosome. Antibody molecules, the classical complement pathway, and the properdin pathway (discussed below) facilitate the first step in the process.

Substances that coat foreign particles and facilitate their recognition and binding to phagocytes are called *opsonins;* the phenomenon is called *opsonization.* Unless the ingested microorganisms have specific mechanisms for preventing the fusion of lysosomes and phagosomes (a process called *degranulation*) to form a *phagolysosome,* the battery of enzymes emptied into the vacuole kills and digests the microorganism.

Complement System. Complement is a group of 11 related proteins normally present in serum. Each of the proteins is designated by convention by the letter C and a number (e.g., C1, C2). Subunits of the proteins are designated by the lower case letters (e.g., C1q, C1r). The system enhances phagocytosis, plays a role in the inflammatory response, and leads to the lysis of foreign cells. It can be activated by two pathways: the classical, or conventional pathway and the alternate, or properdin pathway.

1. Classical (conventional) pathway. The first step in the pathway is the activation of the system. Usually this takes place when the first component of complement (C1) reacts with antigen-antibody complexes. A sequence of events occurs after C1q and C1r bind to the antigen-antibody complex to make C1s enzymatically active (designated $\overline{C1s}$). The $\overline{C1s}$ cleaves C4 into two subunits (C4a and C4b), the start of a number of steps which lead to the assembly of the *activation unit* which clings to the cell membrane. The final steps in the series involve the cleavage of C5 and the binding of molecules of C5b, C6, C7, C8 and C9 into a complex consisting of 10 molecules which binds to a site on the cell membrane, causing membrane destruction and lysis of the cell.

2. *The Properdin (Alternate) Pathway.* Properdin consists of several noncomplementary serum proteins which can activate complement. The properdin pathway does not require the presence of specific antibodies for initiation and therefore does not utilize C1, C2, or C4, all complement components involved in the early steps of the classical pathway. Components of foreign cells react with the properdin components to form active C3 and C5 convertases which then activate the formation of the attack complex.

Interferon. Interferon is a soluble protein released by virus-infected cells which serves as an inhibitor for further virus replication. Interferon is specific for the host (humans, chickens, etc.) but not specific for any particular virus. The action of interferon may be to block the transcription or translation of viral mRNA.

Acquired Resistance (Immunity). Vertebrates, particularly higher vertebrates, possess a highly developed, sophisticated system for developing resistance to particular species and varieties.

SYSTEMS OF IMMUNITY. There are two systems of immunity: humoral and cell-mediated.

Humoral Immunity. The immune response in the humoral system of immunity is mediated by immunoglobulins, which are serum proteins with antibody activity produced by lymphocytes called B cells. This systems protects against bacterial infections and viral reinfections.

Cell-mediated Immunity. The immune response here is mediated by live cells called T cells which have become sensitized to antigens (foreign substances). (The term cellular immunity is also used for the general process of phagocytosis in host defense against disease.) This system protects against fungi, protozoa, and viruses and plays a role in the rejection of tumor cells and foreign tissues.

ANTIGENS. Antigens, (Ag) are substances or parts of cells that gain entrance to the body and elicit a specific immunological response. Antigens have the following characteristics:

1. They are foreign to the host's immunological system, that is, their molecules possess *antigenic determinants,* particular surface configurations that are not present in the molecules of the host's tissues, that trigger immunological responses.

2. They are macromolecules (mol wt > 6000 daltons). These can be proteins, polysaccharides, glycoproteins, nucleoproteins, and glycolipids. Often they are structural parts of the capsules, capsids, cell walls, cell membranes, and flagella of bacteria, viruses, protozoa, and other microorganisms. Smaller molecules with appropriate characteristics known as *haptens* become antigenic when bound to larger carrier molecules.

3. Native or undenatured macromolecules are usually better antigens (they elicit stronger antibody response) than denatured or partially degraded antigens.

Antibodies. Antibodies (AB) are specific *immunoglobulins* that are synthesized in response to antigenic stimulation. They are found in the serum portion of the blood. Antibodies have a molecular weight of approximately 150,000 daltons and are composed of four polypeptide subunits; two identical *heavy chains* (50,000 daltons) and two identical *light chains* (25,000 daltons) linked to each other by sulfhydral bonds. The portion of the molecule containing the two light chains and complementary fragments of the heavy chain combine with the antigen and are called *fab* (*f*ragment *a*ntigen *b*inding). The remainder of the molecule containing fragments of the heavy chain is called the *fc* (*f*ragment-*c*rystalized) end. The amino acid sequences at the amino acid terminal ends of both the light and heavy chains of the fab portion of the molecule are variable, while those close to the sulfhydral bridges (hinge region) and the fc end of the heavy chain are more constant. It is believed that the variable portion of the four polypeptides is responsible for the antibody specificity of the molecule by virtue of its spatial configuration which will just fit a particular antigenic determinant group.

Most classes of immunoglobulins have carbohydrate moieties covalently linked to the fc portion of the heavy chains.

CLASSES OF IMMUNOGLOBULINS. Based on differences found in their heavy chains, immunoglobulins fall into five distinctive classes: IgA, IgG, IgD, IgE, IgM. The heavy chains corresponding to each have been designated by Greek letters : α (alpha), γ (gamma), δ (delta), ε (epsilon), μ (mu). Two major classes of light chains, κ (kappa) and λ (lambda), have been described. Subclasses based on antigenic differences (IgG1, IgG2, IgG3, etc.) have also been designated. The amino acid sequences of the constant regions determine the class or biological role of the immunoglobulins (see below) while the variable regions of the molecule determine specificity.

1. *IgG*. About 85% of the immunoglobulins found in serum belong to this class. They are also found extravascularly, particularly beneath the epithelium. This class crosses the placental barrier, thereby affording newborn infants temporary protection against infectious diseases for which the mother has generated antibodies. IgG immunoglobulins neutralize bacterial toxins, inactivate extracellular bacteria and viruses, and bind to microorganisms, thus making them more susceptible to phagocytosis.

2. *IgM* (macroglobulin) constitutes about 6% of the total immunoglobulin. The molecular weight of IgM is approximately 900,000, the result of five of the four-chain basic units being attached to each other by disulfide bridges near the hinge areas of the heavy chains. Confined largely to the bloodstream, IgM are functionally similar to IgG antibodies.

3. *IgA* constitutes about 10% of the immunoglobulins. Although around 40% of the IgA is present in the bloodstream in a monomeric form, the important fraction, *secretory IgA* (sIgA), is found in secretions on epithelial tissue (e.g., saliva, tears, colostrum, mucus in intestinal, respiratory, and genital tracts). The sIgA fraction is a dimer consisting of two molecules of IgA joined by a polypeptide chain (J chain) and a polypeptide transport piece. It is resistant to enzymatic digestion and functions as a local immune system that protects mucosal surfaces from invasion by pathogens.

4. *IgD* is only a minute fraction of the total body immunoglobulins. They are not yet well characterized.

5. *IgE* (atopic antibody; skin-sensitizing antibody; anaphylactic antibody) also constitutes a minute fraction ($\sim 0.002\%$) of the total immunoglobulins in normal human serum. The principal signifi-

cance of this class is its role in severe, acute, and occasionally fatal allergic reactions. The fc portion of IgE antibodies has an affinity for most cells and basophils that contain histamine. When released, histamine and related physiologically active substances exert a number of effects which may be deleterious.

TYPES OF IMMUNITY. Immunity is acquired in a variety of ways. There are two general types of immunity: active and passive, each of which can be natural or artificial.

Active Immunity. Active immunity is the result of an individual's own response to a challenge by an antigen. The immune response can be mediated in two ways: through circulating antibodies (*humoral immunity*) or through activated living cells (*cell-mediated immunity*). Active immunity can be naturally acquired through infection or artifically induced through the injection of immunizing substances: vaccines and toxoids.

Vaccines. Vaccines are suspensions of microorganisms that have been killed or attenuated.

Killed Microorganisms. Microorganisms are killed through chemical or heat treatment. Pertussis, cholera, and typhoid fever vaccines are suspensions of killed bacteria. The Salk poliomyelitis vaccine consists of killed viruses.

Attenuated, Living Microorganisms. Microorgansims are said to be attenuated when their virulence has been reduced or lost through special treatment or through mutation. The vaccines for measles, mumps, and oral polio (Sabin) contain attenuated viruses.

Toxoids. Toxoids are chemically denatured or modified toxins. Tetanus and diphtheria exotoxins are denatured by heat and formalin.

Passive Immunity. Passive immunity is a temporary (weeks or months) immune state acquired by the transfer of antibodies from a person or animal that has manufactured them to another individual. It is said to be natural when the antibodies are transmitted from mother to fetus through the placenta or to the newborn through colostrum, and artificial when antibodies are received from the injection of antiserum. In artificial passive immunity, the antibodies are transferred in serum; the immunizing material is thus known as *antiserum,* or if it is directed against a toxin, as *antitoxin.*

ANTIGEN-ANTIBODY REACTIONS. *Serology* is the formal name of the scientific subdiscipline that studies antigen-antibody reactions. Information derived from serological studies is extremely important in clinical diagnostic microbiology. A number of different types of antigen-antibody reactions have been recognized.

Neutralization. Microbial toxins are neutralized when antibodies combine with them in such a way as to block the activity of the toxin. An antibody that neutralizes a toxin is known as an *antitoxin*. Neutralizing reactions between antibodies and the surface antigens of viruses prevent adsorption of viruses to their hosts.

Precipitation. A combination of soluble antigens with more than one combining site (polyvalent) and antibodies with more than one combining site (bivalent or multivalent) forms large complex masses or lattices of joined molecules which precipitate in readily visible flocs. Relatively equal concentrations of both the antibodies and antigens are necessary for precipitation to take place. If antigens are present in excess then they will saturate all the combining sites on the relatively few antibodies and lattices will not form. The same situation holds if antibodies are in excess. A common serological technique (Ouchterlony) used for the demonstration of similarities and partial identities of antigens takes advantage of precipitation reactions between antigens and antibodies. In the Ouchterlony technique the antigen and antiserum to the antigen are placed in small wells in agar only a few millimeters apart. Both diffuse into the agar. When they meet in the appropriate concentrations, they form a precipitin band or bands if more than one antigen and antibody are present.

Agglutination. When the antigens are on the surfaces of cells, antigen-antibody reactions with bivalent or multivalent antibodies lead to a clumping or agglutination of the cells. Cells that are aggregated by agglutinins or *opsonized* by them are more readily taken in by phagocytes.

Lysis. Lysis, or rupture, of microorganisms can take place in antigen-antibody reactions with whole cells. Such reactions involve the complement system described earlier.

Immobilization. Antibodies directed against bacterial flagella produce immobilization.

Fluorescent Antibodies. Fluorescent organic compounds such as rhodamine B or fluorescein isothiocyanate can be attached to antibody molecules without altering the antibody's specificity for antigens. Large antigens such as bacteria can be visualized and thus identified through a fluorescent microscope by virtue of the fluorescent-tagged antibodies on their surfaces.

Complement Fixation. During certain types of antigen-antibody reactions, fractions of complement are altered in such a way that they are no longer available to participate in new reactions. A common serological test, the complement fixation test, takes advantage of this knowledge. The purpose of the complement fixation test is to determine whether specific bacterial antibodies are present in a serum sample. A known antigen and a known amount of complement are added to the serum being tested. (Before being used in the test, the serum has been heated to 56°C to inactivate the complement in the serum.) The antigen-complement-serum mixture is incubated to allow an antibody-antigen reaction to take place. An indicator system (hemolytic system) is then added to the mixture to see if the complement has been fixed (used up). The indicator system consists of sheep red blood cells and an antibody made by rabbits against sheep red blood cells. In a *positive* complement fixation test, the test serum has antibodies against the antigen which use up the complement and make it unavailable for use by the indicator system. In a *negative* complement fixation test, there are no antibodies in the test serum against the antigen; the complement is not used up and is available when the indicator system is added. The antisheep red blood cell antibody binds with the sheep red blood cells and complement, causing their lysis. Thus, if there is no red blood cell lysis, the serum sample contains the antibodies that are being tested.

ANTIBODY PRODUCTION. Long before the cellular and molecular aspects of antibody formation were being unraveled, it was established that there was a latent period, or lapse of time, between the initial challenge to an animal by the injection of an antigen (*primary stimulus, primary response*) and the appearance of antibody in circulation. The amount of antibody produced after a primary response gradually increases and then falls off. Response to a second or booster dose of the same antigen is much more rapid and the amount of antibody produced in a *secondary,* or *anamnestic reaction,* is 10 to 100 times that produced after a primary stimulus.

The cells that actually produce immunoglobulins, called *plasma cells,* are found in the spleen and lymph nodes. Three types of cells, macrophages, T cells, and B cells, are actually involved in a sophisticated system of cellular interactions which lead to antibody production. Macrophages initiate the process by nonspecifically phagocytizing the antigens. Antigenic materials become recognizable on the surface of the macrophages. Next, B and T cells combine with the macrophages to form a three-cell complex. T cells become active and release substances that stimulate B cells (now called plasma cells) to produce antibody. (B cells can, however, make IgM antibodies without T cell stimulus.)

T Cells. T cells are formed in the bone marrow and mature in the thymus. Evidence suggests that they are long-lived (months to years). They are mobile and proliferate upon antigen stimulation. They do not themselves produce immunoglobulins, or have them on their surfaces, but they stimulate B cells to do so. Once sensitized to a particular antigen, they kill cells with that antigen and also incite macrophages to attack them (cell-mediated immunity). They also produce lymphokines, substances that cause inflammation.

B Cells. B cells also originate in the bone marrow but they mature in the lymphoid tissues in mammals or the *bursa of Fabricus* in birds. They have relatively short lives (days to weeks) and they are relatively immobile. Complement receptors and immunoglobulins are found on their surfaces. They also proliferate upon antigenic stimulation and are the cells that synthesize immunoglobulins.

Chapter 12

SUMMARY OF PROCARYOTIC GROUPS

Ideally, organisms are classified on the basis of natural, or evolutionary, relationships. Such relationships are not always clear. Taxonomists working with procaryotes are particularly frustrated because there is not yet sufficient evidence to permit the construction of classification schemes that will encompass all procaryotic groups and reflect accurately their phylogenetic relationships. Recognizing this problem, the editorial board for the eighth edition of *Bergey's Manual of Determinative Bacteriology*, the standard reference source for bacteriologists, decided to divide the procaryotes into 19 parts, or hierarchies, based on a few readily determinable criteria (table 12.1).

TAXONOMIC STATUS OF THE PARTS OF *BERGEY'S MANUAL*

"Part" does not directly correspond to any of the traditional taxonomic terms (class, subclass, order, etc.). Usually, organisms comprising each part seem to be naturally (presumably evolutionarily) related to each other. Many of the parts contain natural assemblages of organisms that reasonably can be organized into taxonomic units at the ordinal (order) and family levels. However, some parts include organisms that are *incertae sedis,* that is, their relationship to other procaryotes is uncertain. They are placed in a particular part because, on the basis of available evidence, they *seem* to have greater affinity for that hierarchy than for any other.

Table 12.1. The Nineteen Categories of Procaryote Classification from Bergey's Manual of Determinative Bacteriology

Part 1.	Phototrophic Bacteria
Part 2.	Gliding Bacteria
Part 3.	Sheathed Bacteria
Part 4.	Budding and/or Appendaged Bacteria
Part 5.	Spirochetes
Part 6.	Spiral and Curved Bacteria
Part 7.	Gram-Negative Aerobic Rods and Cocci
Part 8.	Gram-Negative Facultatively Anaerobic Rods
Part 9.	Gram-Negative Anaerobic Bacteria
Part 10.	Gram-Negative Cocci and Coccobacilli
Part 11.	Gram-Negative Anaerobic Cocci
Part 12.	Gram-Negative Chemolithotrophic Bacteria
Part 13.	Methane-Producing Bacteria
Part 14.	Gram-Positive Cocci
Part 15.	Endospore-Forming Rods and Cocci
Part 16.	Gram-Positive Asporogenous Rod-Shaped Bacteria
Part 17.	Actinomycetes and Related Organisms
Part 18.	Rickettsias
Part 19.	Mycoplasmas

CRITERIA USED TO SEPARATE AND IDENTIFY MAJOR GROUPS OF BACTERIA

Gram-staining characteristics and morphology are two of the major properties that are used to separate many of the major groups (parts) of procaryotes. Besides general form (rods, cocci, spiral, etc.), morphological characteristics used for separation include the absence of a cell wall (in mycoplasmas), possession of a system of axial fibers (in spirochetes), growth within sheaths (in sheathed bacteria), formation of endospores (in Bacillaceae), formation of filamentous or coryneform cells and colonies (in Actinomycetes), and the formation of buds and stalks (in Part 4). The rickettsia are separated from other major groups on the basis of their obligate parasitism of warm-blooded hosts and insects. Other major groups (e.g., phototrophic, Part 1, and chemolithotrophic, Part 12) are separated from all other procaryotic groups on the basis of their energy-generating systems.

IDENTIFICATION OF AN UNKNOWN BACTERIUM

The procedures to be followed for the identification of bacteria are briefly outlined by S.T. Cowan and J. Liston in an introductory section of *Bergey's Manual*. They are as follows:

1. Isolate an axenic (pure) culture.
2. Determine its energy requirements (phototrophic, chemolithotrophic, chemoorganotrophic, etc.)
3. Gram-stain the cells and examine them by conventional light microscopy; examine living cells by phase contrast. These procedures yield information on the cell wall characteristics (gram + or −), general cell morphology, (rod, coccus, vibrio, spiral, spirochete, filament, etc.), the possession of spores, stalks, sheaths, and so on, and the type of motility (gliding or flagellated).
4. If critical, stain flagella and examine in the light and/or electron microscope.
5. Examine colonies for growth characteristics and products that might be released from them (e.g., pigments).
6. Test the culture's ability to grow with or without air (aerobe, anaerobe).
7. Determine the oxidative or fermentative abilities of heterotrophs, the inorganic energy sources of chemolithotrophs, and the ability of phototrophs to grow in the dark in suitable media.
8. After identifying the organism as a member of a major group, inoculate diagnostic media with it to determine its specific physiological and biochemical characteristics. Such tests indicate the presence or absence of particular metabolic pathways or enzymes. Among the more common diagnostic tests are those that determine abilities to ferment various carbohydrates, decarboxylate or deaminate amino acids, produce indole from tryptophan, reduce nitrate, split urea, rapidly decompose H_2O_2, coagulate blood plasma, liquefy gelatin, and produce H_2S during the decomposition of sulfur-containing amino acids.

A standard reference that gives precise directions for carrying out the various diagnostic tests used in clinical microbiology is *Manual of Clinical Microbiology* (published by the American Society for Microbiology). A very useful key to the genera of bacteria by V. Skerman is widely available. Skerman is also the author of a manual, *Abstracts of Microbiological Methods,* which summarizes most diagnostic tests and gives references to published papers where more details can be found.

PHOTOTROPHIC (PURPLE AND GREEN) BACTERIA: BERGEY'S MANUAL, PART 1

This group of purple and green phototrophic bacteria includes a wide variety of morphological (cocci, rods, spirals, budding) and locomotor (nonmotile, polarly and peritrichously flagellated, gliding) types. They are common in lakes and salt marshes. Bacteriochlorophylls and carotenoid pigments give them their characteristic purple, green, brown, pink, rust, and orange colors. The photosynthetic pigments of the purple bacteria are associated with tubes or lamellae which ramify from, and are continuous with, the plasma membrane. In the green bacteria, pigments are associated with cylindrical vesicles (*chlorobium* vesicles) underlying but not continuous with the cell membrane.

Whereas in cyanobacteria, algae, and higher plants, photosynthesis is primarily *oxygenic*, that is, O_2 is produced, in the purple and green bacteria it is *anoxygenic* (O_2 is not produced) and only one light reaction (photosystem I) is involved. Moreover, in green and purple bacteria, water is not the reductant; instead, reducing power is obtained from H_2, reduced sulfur compounds such as H_2S, or organic compounds. Although purple and green bacteria grow photosynthetically under anaerobic conditions, some are capable of growing chemoorganotrophically in the dark under aerobic conditions, or in the presence of suitable electron acceptors (e.g., NO_3 and S^0).

Four families of phototrophic bacteria are now recognized: Rhodospirillaceae, Chromatiaceae, Chlorobiaceae, and Chloroflexaceae, the latter not listed in *Bergey's*. The first three constitute the single order Rhodospirillales in *Bergey's*.

Rhodospirillaceae: Purple Nonsulfur Bacteria. These photoorganotrophic purple bacteria differ from other photosynthetic groups in their inability to oxidize elemental sulfur to sulfate (they can, however, oxidize sulfide to sulfate) and their ability to grow aerobically in the dark. Some species are also fermentative in the dark. Methanol, formate, and isopropyl alcohol are common electron donors. The family is morphologically diverse including as it does species of the polarly flagellated *Rhodospirillum* and *Rhodomicrobium*, a genus of peritrichously flagellated budding bacteria.

Chromatiaceae: Purple Sulfur Bacteria. The bacteria in this family are able to oxidize elemental sulfur to sulfate, and their cells are often filled with sulfur granules. The Chromatiaceae are more lim-

Phototrophic (Purple and Green) Bacteria

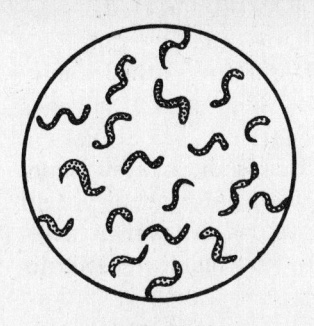

Rhodospirillum
∼ X 2,000

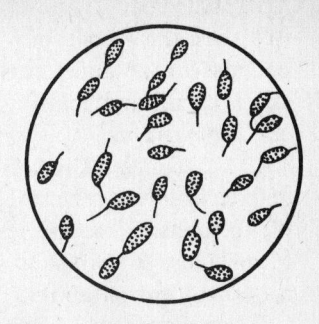

Rhodomicrobium

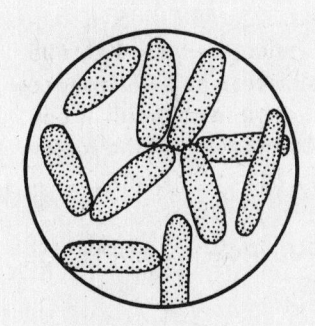

Synechococcus

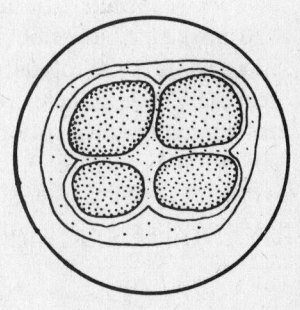

Gloeocapsa (formerly *Chroococcus*)
∼ X 400

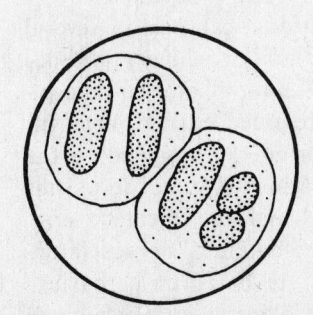

Gloeocapsa
∼ X 1,400

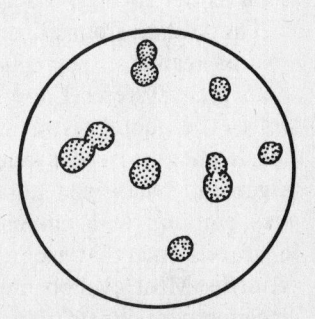

Chamaesiphon
∼ X 1,500

ited than the Rhodospirillaceae in their ability to use organic compounds for photoheterotrophic growth. This family also includes cocci, bacilli and polarly flagellated spirillae. The best-known genus is *Chromatium*.

Chlorobiaceae: Green Sulfur Bacteria. This is the largest family of green photosynthetic bacteria. Many are obligate anaerobes and phototrophs. Some species form gas vacuoles and live planktonically in lakes. Rods, spheres, branching rods, and filamentous forms are found in the family.

Chloroflexaceae: Gliding Green Bacteria. The Chloroflexaceae are morphologically similar to the cyanobacteria, but they lack photosystem II and related metabolic abilities to carry out oxygenic photosynthesis.

Rhodospirillaceae and Chromatiaceae contain bacteriochlorophyll a (characteristic absorption maxima 375, 590, 805, 830 to 890 nm) and bacteriochlorophyll b (absorption maxima 400, 605, 835 to 850, 1020 to 1040 nm). The Chlorobiaceae and Chloroflexaceae have either bacteriochlorophyll c (chlorobium chlorophyll 660, absorption maxima 745 to 755 nm) or bacteriochlorophyll c (chlorobium chlorophyll 650, absorption maxima 705 to 740 nm).

CYANOBACTERIA (NOT LISTED IN *BERGEY'S MANUAL*, 8TH EDITION)

Cyanobacteria are one of the largest subgroups of gram-negative procaryotes. They are common in freshwater, many marine habitats, and in soils and rocks. Most interestingly, cyanobacteria are abundant or dominant in several types of extreme environments such as hot springs, desert soils, and saline lakes ("Sabkhas").

They differ from all other photosynthetic procaryotes in several related respects: their photosynthetic metabolic machinery includes both photosystems I and II; reducing power for photosynthesis involves the photolysis of water; and O_2 is released during photosynthesis. Because the cyanobacteria were originally assigned to the algae (as blue-green algae), the classification of these organisms was built up with emphasis on structural and ecological criteria, characteristics that are important under the botanical nomenclatural code. Most cyanobacterial descriptions are based on herbarium specimens or illustration and *not* on physiological or biochemical properties of cultures.

Cyanobacteria possess only one kind of chlorophyll, chlorophyll a. The molecule can exist in different forms determined by the chemical environment and distinguishable by their absorption spectra (maximum absorption at $\sim$670, 680 or 700 nm). They also have blue, water-soluble pigments, phycobiliproteins, phycocyanin (maximum absorption 620 nm), allophycocyanin (maximum absorption 650 nm), and allophycocyanin B (maximum absorption 670 nm). In addition, some cyanobacteria have phycoerythrin (maximum absorption 550 to 570 nm), phycoerythrocyanin (maximum absorption 565 nm, secondary peak at 590 nm), and other phycoerythrins.

Recently (Rippka and colleagues, 1979), a group of researchers at the Institut Pasteur have examined a large number of strains of cyanobacteria and have suggested newer approaches to the taxonomy of the group. Major differences in structure and development are used to divide the cyanobacteria into two major divisions which are further subdivided into a total of five major subgroups or sections. Cyanobacteria in sections I and II are unicellular or colonial aggregates held together by additional outer cell wall layers. Cyanobacteria assigned to section I reproduce by binary fission or by budding; those in section II reproduce by multiple fission, giving rise to small daughter cells (baeocytes), or by both multiple fission and binary fission.

Section I. This section contains some of the genera from the orders Chorococcales and Chamaesiphonales of the traditional botanical classification. Included is a relatively new genus *Gloeobacter* which has an ultrastructure unlike that of any other cyanobacterium. Species of this genus lack thylakoids and typical phycobilisomes; instead, phycobiliproteins are located on the inner surface of the cell membrane. The genera *Synechococcus* and *Gloeothece* divide only in one plane, *Synechocystis* and *Gloeocapsa* in two or three planes. In both pairs, the latter genus is ensheathed while the former is not. *Chamaesiphon* is a genus of budding unicellular cyanobacteria.

Section II. Section II includes many of the remaining genera of the botanical order Chamaesiphonales. Two genera, *Dermocarpa* and *Xenococcus,* reproduce only by multiple fission. The baeocytes (small reproductive cells) of *Xenococcus* have a fibrous outer wall layer and do not glide, while those of *Dermocarpa* contain only peptidoglycan and outer membrane layers. In *Dermocarpella,* bina-

176 Summary of Procaryotic Groups

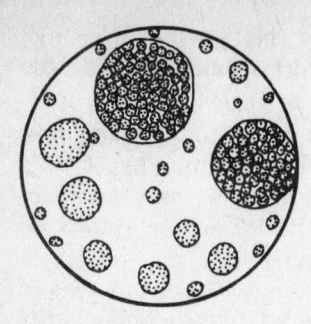

Dermocarpa
~ X 1,000

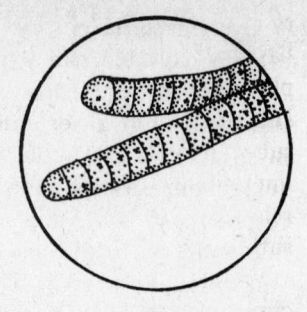

Oscillatoria
~ X 800

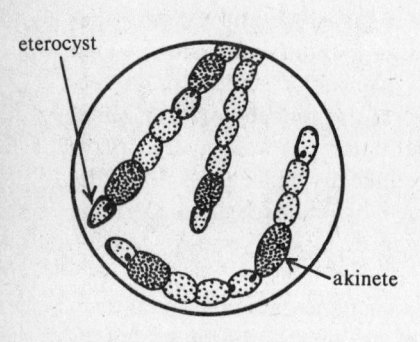

Anabaena
~ X 800

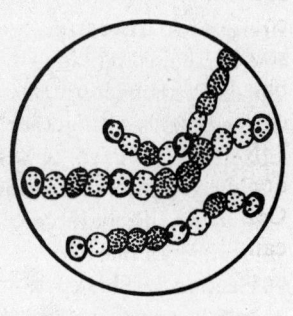

eterocyst
akinete

Nostoc
~ X 800

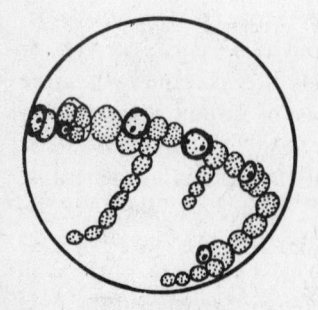

Fischerella
~ X 1,500

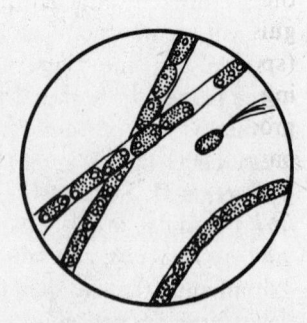

Sphaerotilus natans
~ X 1,500

ry fission leads to pear-shaped structures composed of one or two basal cells and one apical cell which subsequently undergoes multiple fission to yield motile caeocytes. In *Myxosarcina* and *Chroococcidiopsis,* binary fission yields cubical cellular aggregates which subsequently undergo multiple fission. The latter genus forms only immobile baeocytes with fibrous outer walls. A group of organisms related to *Pleurocapsa* form irregular aggregates by binary fission; subsequent multiple fission yields motile baeocytes.

Section III. Assigned to section III are many of the nonheterocystous, filamentous organisms from the botanical order *Nostocales.* Microbiologists have found that sheath formation among these organisms is quite variable in culture and therefore not a reliable morphological property. *Spirulina* is a well-known helical corkscrew genus. Its thinly sheathed, motile filaments (trichomes) are composed of cylindrical or disk-shaped cells; there is little or no constriction between adjacent cells. All the other genera have straight trichomes. Another common, thinly sheathed blue-green bacterium, *Oscillatoria,* has similar cells not separated by deep constrictions.

Many of the species of *Lyngbya, Phormidium,* and *Plectonema* are similar except they are immotile and enclosed by heavy sheaths. Cells of *Pseudoanabaena* are cylindrical so that deep constrictions can be seen between adjacent cells. The trichome is motile and not ensheathed.

Section IV. These are filamentous, heterocystous cyanobacteria that divide in only one plane. Botanists traditionally include these genera in the order Nostocales. In *Anabaena, Nodularia,* and *Cylindrospermum* reproduction occurs by random fragmentation of the vegetative trichomes into shorter cell chains that are indistinguishable from the parents. These organisms can also form akinetes (spores) with thickened outer walls that protect the cell against drying or freezing. If akinetes (spores) are formed, they germinate to produce short filaments that are structurally indistinguishable from parental trichomes. Heterocysts, heavy walled cysts in which nitrogen fixation takes place, are formed at both ends of the trichome in *Cylindrospermum* and terminally and within the trichome in *Anabaena* and *Nodularia.* Cells in *Anabaena* are spherical, ovoid, or cylindrical, whereas those of *Nodularia* are disk-shaped. Some of the other genera in Section IV, such as *Nostoc, Scytonema,* and

Calothrix, form hormogonia that differ from mature trichomes by the absence of heterocysts and other differentiated cell types.

Section V. The filamentous, heterocystous cyanobacteria in this section divide in more than one plane. These organisms, traditionally placed in the order Stigonematales, reproduce by random trichome breakage, by formation of hormogonia, and by germination of akinetes. *Fischerella* and *Chlorogloeopsis* are two well-known genera that belong to this section.

GLIDING BACTERIA: *BERGEY'S MANUAL,* PART 2

These bacteria have no flagella, but move by gliding, leaving slime tracks along surfaces over which they have moved. They are morphologically quite diverse, containing rods, spirillae, and filamentous forms. Two major subdivisions, given the rank of order, are recognized: Myxobacterales and Cytophagales.

Myxobacterales. The myxobacters, or slime bacteria, are short rods that lack a rigid cell wall. Under appropriate environmental conditions (e.g., exhaustion of the media), the individual cells swarm together and form *fruiting bodies.* Some of the cells within the stalked fruiting body transform into resistant resting structures called *myxospores.* In some genera the myxospore is enclosed in a hard slime capsule. These encapulated structures, known as *microcysts,* probably protect the organisms from desiccation when their habitat is subjected to drying. Larger walled structures, called *cysts,* enclose the myxospores of other genera. Myxospores are more resistant to drying and heat than are vegetative cells, but not as resistant as *Bacillus* endospores. Myxobacters are common in soils, compost, rotting wood, and manure.

Cytophagales. Bacteria belonging to the Cytophagales play important roles in various biogeochemical mineral cycles. *Cytophaga* and related genera are important in the decomposition of many natural structural macromolecules (e.g., cellulose and chitin). *Thiothrix,* an organism commonly found in the activated sludge systems in sewage plants, oxidizes H_2S. Though some genera of the Cytophagales form microcysts, they do not form them on fruiting bodies. *Beggiatoa,* an organism that resembles *Oscillatoria,* forms mats in mud layers of lakes, decaying seaweeds, sewage, and other habitats rich in H_2S. *Beggiatoa* can use H_2S as an energy source which it converts to sulfur granules and then to sulfate.

SHEATHED BACTERIA: *BERGEY'S MANUAL*, PART 3

Sheathed bacteria are found in slowly running fresh water, stagnant pools, and waters polluted with wastes from paper or dairy industries, sewage, and other habitats rich in organic matter. As their name suggests, vegetative growth takes place within long tubes or sheaths, sometimes as long as 1 cm. The sheaths of many genera in this group become encrusted with iron or manganese oxides. Several genera, *Sphaerotilus* and *Leptothrix,* produce polarly flagellated cells (swarmers) at the open end or breaks in the sheath wall. Electron microscopic observations have shown that the polar flagella are really tufts of intertwined flagella (lophotrichous) that give the appearance of a single large flagellum. Common species such as *Sphaerotilus natans* and *Leptothrix ochracea,* which have yellow-brown, iron-containing sheaths, are sometimes known as iron bacteria. Cells under appropriate environmental conditions accumulate large quantities of poly-β-hydroxybutyrate which forms globular inclusions.

BUDDING AND/OR APPENDAGED BACTERIA: *BERGEY'S MANUAL*, PART 4

Members of this group form slender stalks, hyphae, or appendages and grow from one end of the cell (polar growth). When they divide, most genera produce daughter cells that are unequal in size. Slender extensions of cells that contain cytoplasm and are bounded by the cell wall are known as *prosthecae,* and the bacteria that have them are called prosthecate bacteria (e.g., *Hyphomicrobium, Hyphomonas, Pedomicrobium, Caulobacter, Prosthecomicrobium, Asticcacaulis,* and *Thiodendron*). Some genera (e.g., *Gallionella* and *Nevskia*) have stalks that are elongated, excreted extensions from the slime layer outside the cell wall, while in other genera (e.g., *Caulobacter*) the stalk is rather broad and contains cytoplasm that remains after a daughter cell is budded off from the opposite end of the cell. The daughter cells, or swarmers, of *Caulobacter* have a single polar flagellum. The swarmers eventually settle down on the substrate and form new stalks at the flagella end of the cell, after which the flagellum is lost. Growth and cell division in *Hyphomicrobium* are quite different. A hypha is formed at one end of a cell. As the hypha lengthens, the DNA doubles, and then one

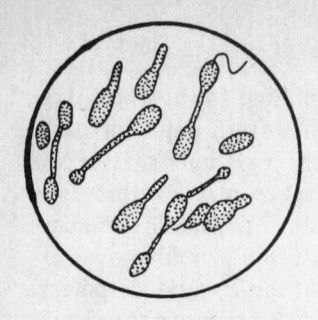

Hyphomicrobium
∼ X 3,500

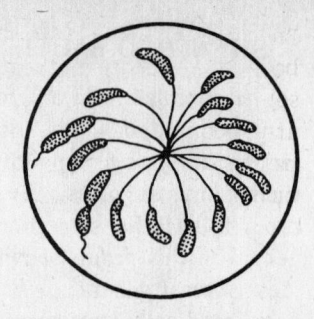

Caulobacter
∼ X 2,000

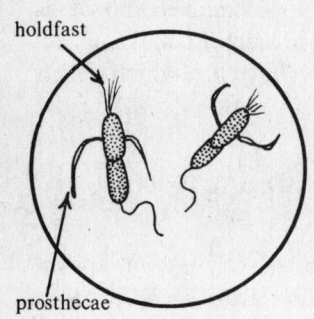

holdfast

prosthecae

Asticcacaulis
∼ X 5,000

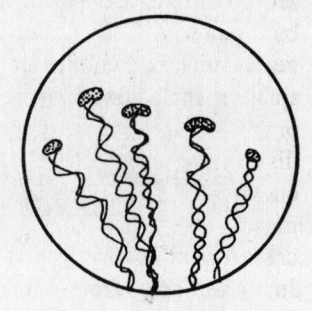

Gallionella minor
∼ X 1,000

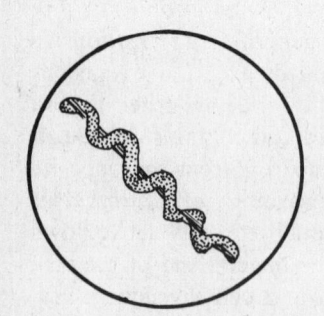

Spirochaeta
∼ X 8,000

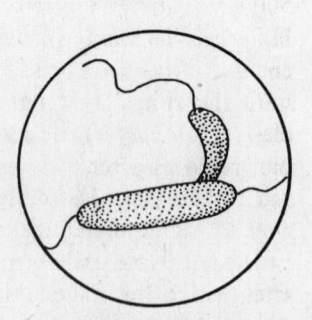

Bdellovibrio attached to *Pseudomonas*
∼ X 8,000

"nuclear body" (genophore) enters the hypha. After the "nuclear body" (genophore) reaches the terminal, a bud forms, which then is separated from the rest of the hypha by a septum. The bud enlarges and a flagellum forms on it before the swarmer separates from the parent. The DNA in the parent can again be replicated and another bud again can form at the end of the hypha.

SPIROCHETES: *BERGEY'S MANUAL,* PART 5

Spirochetes are slender, flexous, helically coiled, unicellular bacteria. They vary in length from a tiny 3 μm to over 500 μm in length. Some species, including the spirochete *Treponema pallidum* that causes syphilis in humans, are so slender ($\sim$0.1 to 0.2 μm) that they are too small to be resolved by the light microscope. They are visualized only by reflected light in a dark-field microscope or by electron microscopy. Some spirochetes occur in chains held together by an outer envelope. Electron microscopic studies show that spirochetes consist of a protoplasmic cylinder intertwined with two or more (up to 100 in *Cristispira*) axial fibers that run between disks at opposite ends of the cell. An outer envelope (outer sheath) encloses both the cylinder and the axial fibers. Many spirochetes move through media with a corkscrew-like motion. Some seem to creep, others flex and lash. Spirochetes divide by transverse fission. In *Bergey's* they are listed as an order (Spirochaetales) with a single family (Spirochaetaceae) divided into five genera on the basis of habitat, pathogenicity, and morphological and physiological characteristics. Members of the free-living genus *Spirochaeta* are very slender (0.2 to 0.75 μm) and are quite variable in length (5 to 500 μm). They are widely distributed in fresh and marine anaerobic H_2S-rich sediments, sewage, and polluted waters. *Cristispira* are endocommensal in the digestive tract of freshwater and marine mollusks. They are commonly found in the crystalline style of clams and oysters. *Treponema* spp. are anaerobic commensal or parasites found in the oral cavity, intestinal tract, and genital regions of humans and animals. They cause syphilis, yaws, and pinta. *Borrelia* spp. are generally parasitic or live on the mucous membranes of humans, rodents, birds, and other animals. Those that cause relapsing fever in humans are transmitted by *Pediculus humanus,* the body louse, and ticks belonging to the genus *Ornithodoros.* The same ticks also transmit diseases to birds (avian spirochetosis) and other animals (e.g., bovine spirochetosis). Infections with *Borrelia* depend upon the feeding habits and preferences of the vector. *Leptospira,* the causative agents of Weils disease in humans, dogs, pigs, and rodents, are the most slender (0.1 µm in diameter) of all the spirochetes. They are usually bent at each end into a semicircular hook. After these aerobic spirochetes enter the body through mucous membranes or breaks in the skin, they multiply, localizing in the kidney and liver and causing nephritis and jaundice. Strains of *Leptospira* have been found seemingly free-living in fresh and saline waters. They could be saprobic or could have entered the waters through the urine of infected animals.

SPIRAL AND CURVED BACTERIA: *BERGEY'S MANUAL,* PART 6

Rigid, helically curved, polarly flagellated rods belonging to the family Spirillaceae, and several other genera (*Bdellovibrio, Pelosigma,* and *Brachyarcus*) that might be related to the family, constitute Part 6 of *Bergey's Manual.* Flagellation may be at one or both poles. Although they are chemoorganotrophs, only a few species in this group ferment carbohydrates. They are found free-living in fresh and marine environments; some are pathogenic. *Spirillum volutans* is a very large cell often used in microbiology and biology classrooms to demonstrate motility. The fascicule (bundle) of about 75 flagella at each pole is easily seen in living cells by phase or dark-field microscopy. *S. volutans* is obligately microaerophilic; it requires oxygen but grows only when the oxygen level is relatively reduced (1 to 9%). Species of *Campylobacter* are found as commensals or pathogens in the oral cavities, intestinal tracts, and reproductive organs of humans and animals. They have a single flagellum at one or both ends. *Aquaspirillum* and *Oceanospirillum* are small-diameter, aerobic, free-living spirillae; the former is found in fresh waters and the latter is obligate (salt-requiring) marine.

Nutritional studies suggest that these genera are adapted for the recycling of organic matter when it is present at very low concentrations. *Bdellovibrio* spp. are small, curved rods which attach to and penetrate the cells of other gram-negative bacteria. They are commonly isolated from soils and from fresh and marine waters. The methods used to isolate *Bdellovibrio* spp. are similar to those

Spiral and Curved Bacteria

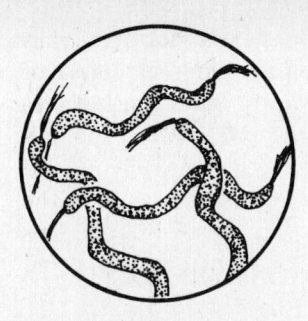

Spirillum volutans
∼ ×1,400

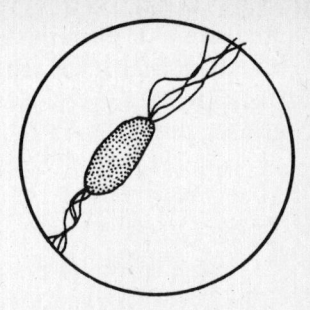

Pseudomonas
∼ ×7,000

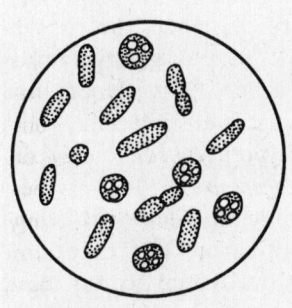

Azotobacter
typical rod-shaped cells
and cysts
∼ ×1,500

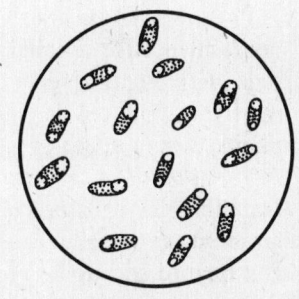

Beijerinckia
showing characteristic
polar lipoid bodies
∼ ×2,000

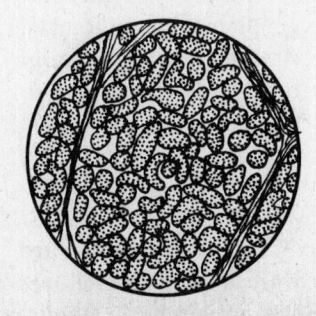

Bacteroids of *Rhizobium*
within clover cell
∼ ×2,500

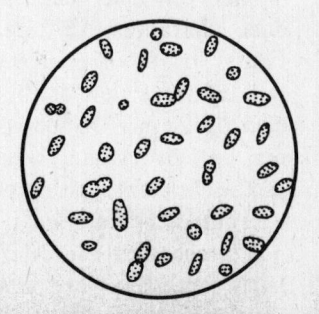

Methylomonas
∼ ×1,500

used to isolate bacterial viruses. A lawn of potential host bacteria is spread on agar plates, then inoculated with the fluid being assayed. The plaques formed by *Bdellovibrio* spp. are analogous to viral plaques. It has been suggested that in nature *Bdellovibrio* spp. play a role in regulating the densities of their hosts.

GRAM-NEGATIVE, AEROBIC RODS AND COCCI: *BERGEY'S MANUAL*, PART 7

A number of families (Pseudomonadaceae, Azotobacteraceae, Rhizobiaceae, Methylomonadaceae, Halobacteriaceae) and a number of genera of uncertain affiliation constitute this part of *Bergey's Manual*.

Pseudomonadaceae. Pseudomonads are chemoorganotrophic, nonfermentative aerobic, straight or curved, motile, gram-negative rods with polar flagella. Species of the genus *Pseudomonas* are widely distributed in soils, aquatic habitats, and in animals and plants, where some are pathogens (e.g., *P. aeruginosa, P. cepacia, P. pseudomallei, P. mallei,* and *P. solanacearum*). Most species, including the parasitic ones, can grow in mineral media with a single organic compound as the sole source of carbon and energy. A few require additional amino acids or vitamins. Acetate can be used as the principal nutrient by all the species that have been characterized. Most species also are able to use lactate, succinate, and glucose. In common with many other gram-negative bacteria, pseudomonads metabolize glucose via the Entner-Doudoroff pathway. Some species and strains produce diffusible blue, yellow, or green pigments that fluoresce in ultraviolet light. Certain species also produce characteristic blue, green, or orange phenazine pigments. Many strains of *P. aeruginosa,* an organism commonly clinically isolated from wounds, burns, and urinary tract infections, can produce "blue pus" which is due to the presence of the blue phenazine pigment, pyocyanin. While pseudomonads are aerobic and use O_2 as their electron acceptor, some can denitrify using nitrate as an alternate acceptor.

Azotobacteraceae. Members of the family Azotobacteraceae are fairly large oval- or rod-shaped cells. They are normal inhabitants of soil, water, and the surfaces of leaves. Some species of *Azotobacter* form cysts which, like endospores, are resistant to desiccation, mechanical stress, ultraviolet and ionizing radiation, and have very low endogenous respiration rates. Vegetative *Azotobacter* cells have the highest respiratory rates measured of any living organisms.

Some species are nonmotile, others are polarly or peritrichously flagellated. All species are strict aerobes, heterotrophic, and capable of growing in nitrogen-free media with organic carbon sources by fixing molecular nitrogen to satisfy their needs for this element. Some representatives are unable to use nitrate as a nitrogen source or grow in peptone-containing media. Most strains are capsulated and produce copious slime. Some strains produce fluorescent pigments. Nonsymbiotic nitrogen fixation by representatives of this family, *Azotobacter, Azomonas, Azotococcus, Beijerinckia,* and *Derxia,* contribute significantly to the fertility of soils. Depending on soil characteristics and related microclimatic conditions, nonsymbiotic bacteria in soils can fix as much as 20 to 50 lb of nitrogen per acre per year.

Rhizobiaceae. The family Rhizobiaceae contains two very well-known genera: *Rhizobium* and *Agrobacterium.* Species of *Rhizobium* characteristically invade root hairs of leguminous plants (family Leguminosae) and stimulate production of root nodules. Within the nodules the normally rod-shaped bacteria transform into pleomorphic (swollen, globular, ellipsoidal, or branched) forms (bacteriods) that are enclosed by plant membranous sacs either singly or in small groups. They fix free nitrogen when in the symbiotic stage. Infectivity and effectiveness in forming nodules varies within wide limits, depending on genetic factors present in both bacterial strains and host plants. Species of *Rhizobium* are common soil inhabitants and can be cultivated on media containing yeast or other plant extracts but may lose their ability to invade or cause nodule formation after prolonged cultivation.

Agrobacterium spp. are also common in soils, mainly those previously contaminated with diseased plant materials. Although *Agrobacterium* spp. do not stimulate root nodule formation on leguminous plants, they cause other types of hypertrophies such as galls, "hairy root" or "woolly knot" disease, and other types of tumors on plants belonging to more than 40 families. The bacteria are intercellular parasites, entering host tissues through pre-existing lesions or abrasions. The ability to cause uncontrolled growths in wounded plants is due to the presence of a large plasmid ($\sim 5\%$ of the total bacterial DNA) in *Agrobacterium* cells, a small portion of which (perhaps 10 to 20 genes) is transferred and transcribed by tumor-induced plant cells.

Methylomonadaceae. Methylotrophic bacteria, that is, bacteria that can use only single-carbon compounds as substrates for growth, are placed in the family Methylomonadaceae. They are aerobes, are widespread in soils and aquatic environments, and can use substrates such as methylamine, trimethylamine, formate, formamide, carbon monoxide, and methanol as well as methane. A specific enzyme system, methane oxygenase enables methylotrophic spp. to oxidize methane. One species, *Methylomonas methanitrificans,* can grow using methane as its source of carbon and energy and molecular nitrogen as its sole N source. Members of the family, which include genera with the prefix methyl- (*Methylomonas, Methylobacter, Methylococcus, Methylosinus, Methylocystis, Methylobacterium,* and *Methanomonas*) are unusual for procaryotes in that they possess relatively large quantities of sterols. Two groups of methane-oxidizing bacteria are recognized. Type I methane-oxidizers have internal membranes arranged as bundles of disk-shaped vesicles distributed throughout the organism and assimilate methane via the ribulose monophosphate pathway. The paired membranes of type II methane-oxidizers are restricted to the periphery of the cell. Type II methane oxidizers assimilate methane via the serine pathway.

Halobacteriaceae. Rods and cocci that require high concentrations of NaCl ($>2M$; $\sim 12\%$) have been placed in the family Halobacteriaceae, a group assigned to this part of *Bergey's Manual* in common with methane-producing bacteria and *Sulfolobus.* The cell walls of halobacteria lack petidoglycan. The cell walls of *Halobacterium* require high concentrations of Na^+ ions for stability, and the phospholipids in their plasma membranes contain long-chain isoprenoids joined by ether linkages to glycerol rather than the more common fatty acids joined by ester linkages. The ribosomes of *Halobacterium* require high concentration of K^+ for stability, and many enzymes require high salt concentrations for activity. The membrane of *Halobacterium* contains patches of bacteriorhodopsin, a reddish-purple pigment that is able to absorb light and generate a proton gradient across the membrane, a photophosphorylation process that leads to the synthesis of ATP. Despite this photophosphorylative ability, halobacteria are typical aerobic heterotrophs. Proteins and proteoses are preferred for growth in complex media. *Halobacterium* spp. have a tuft of polar flagella; *Halococcus* is nonmotile. Both genera are in abundance in salt lakes, salterns, the Dead Sea, and in places where salted fish are stored.

Gram-Negative, Aerobic Rods and Cocci

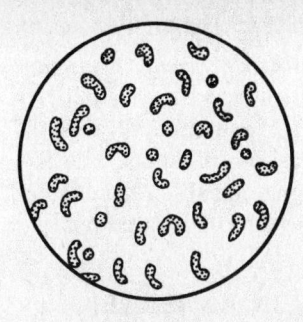

Methylocystis
~ ×1,500

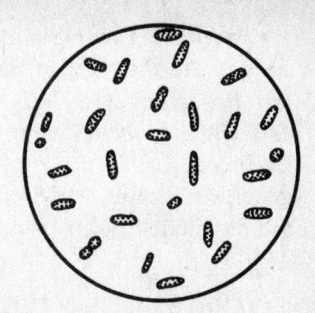

Salmonella typhi
~ ×2,000

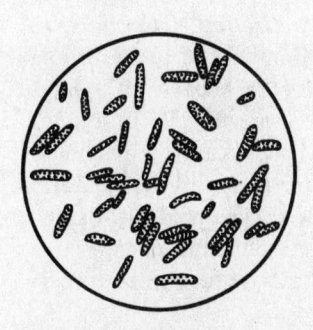

Yersinia pestis
~ ×2,000

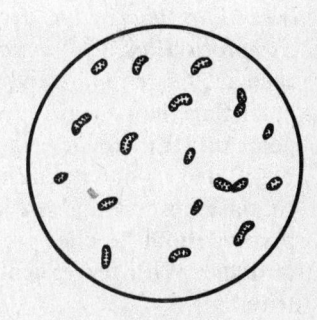

Vibrio cholerae
~ ×2,000

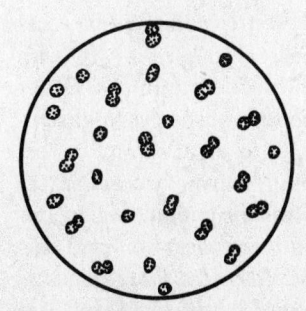

Neisseria gonorrhoeae
~ ×2,000

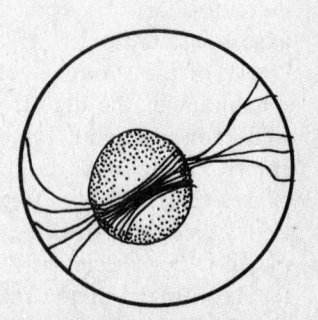

Nitrosococcus oceanus showing cytomembranes and flagella
~ ×10,000

Genera of Uncertain Affiliation. Among the genera tentatively assigned to this part of *Bergey's Manual* are *Alcaligens, Acetobacter, Brucella, Bordetella, Francisella,* and *Thermus*. *Acetobacter* and *Gluconobacter* carry out an incomplete oxidation of alcohols leading to the formation of organic acids as end products. The two genera of acetic acid bacteria differ morphologically. *Acetobacter* are flagellated peritrichously; *Gluconobacter* are flagellated polarly.

GRAM-NEGATIVE, FACULTATIVELY ANAEROBIC RODS: *BERGEY'S MANUAL,* PART 8

This group includes perhaps the best-known family of bacteria, the Enterobacteriaceae, and the family Vibrionaceae plus a few other genera of uncertain affiliation (*Zymomonas, Chromobacterium, Flavobacterium, Haemophilus, Pasteurella, Actinobacillus, Streptobacillus, Calymmatobacterium*). Members of the Vibrionaceae are curved or straight rods that are motile because of a single polar flagellum. Some species have two or more flagella in a single polar tuft. Enterobacteriaceae, in contrast, are small, peritrichously flagellated or nonmotile. Enterobacteria are generally fermentative catalase positive and oxidase negative, whereas vibrios are oxidase positive. Both families are facultatively anaerobic and chemoorganotrophic. With few exceptions, enterobacteria reduce nitrate to nitrite.

Enterobacteriaceae. A battery of well-known differential media and tests are used to classify the enterobacteria. These include the ability to ferment glucose and lactose with the production of acid and/or gas, utilization of particular carbon sources (e.g., citrate, mannitol, tartrate, acetate), tests for urease, phenylalanine deaminase, degradation of tryptophan, and H_2S production. The best known of the enterobacteria is, of course, *Escherichia coli*. It is an inhabitant of the digestive tracts of warm-blooded animals. Other well-known genera, *Shigella* and *Salmonella,* cause severe gastroenteritis (bacterial dysentery) and typhoid fever. *Proteus* rarely causes enteritis but is frequently responsible for urinary tract infections. *Klebsiella pneumonia* is rarely involved in pneumonia in humans. Other species of *Klebsiella* commonly found in soil and water fix nitrogen under anaerobic conditions. *Serratia marcescens* is commonly used in introductory microbiology laboratory exercises because many strains form red or magenta colonies. Species of *Er-*

winia are associated with plants as epiphytes, saprophytes, or as pathogens causing necrotic wilt disease (e.g., fireblight, corn wilt) and various rotting diseases (e.g., onion rot, pineapple rot, potato storage rot, potato black leg). *Yersinia pestis* is the causative organism of plague in humans, rats, ground squirrels, and other rodents.

Vibrionaceae. Vibrionaceae also have fermentative and respiratory metabolisms. Most species are found in aquatic habitats. Some species are well known. *Vibrio cholerae* and *V. parahaemolyticus* are common in waters near human populations. *V. cholerae* causes cholera, once a common and dreaded intestinal disease. *V. parahaemolyticus*, commonly isolated from sewage-contaminated estuarine habitats, also causes enteritis when it is ingested with raw shellfish and finfish. *Aeromonas hydrophila* is well known to biologists as the agent that causes red leg disease in frogs. *Beneckea* and *Photobacterium* spp. are found in the luminescent organs of marine fish (e.g., flashlight fish) and grow as saprophytes on the surfaces of dead fish.

Genera of Uncertain Affiliation. *Chromobacterium violaceum* occasionally causes serious pyogenic or septicemic infections of humans and other animals. *Haemophilus influenzae* is a common inhabitant of the nasopharynx of humans and is isolated in clinical conditions from paranasal sinuses, sputum, conjunctivae, and various other sites. *Pasteurella multocida* is a pathogen causing hemorrhagic septicemia in birds, mice, rabbits, and Asian and African ruminants and pneumonic conditions in cattle, sheep, and pigs. *Streptobacillus moniliformis*, an inhabitant of the nasopharynx of wild and laboratory rats, causes streptobacillary rat-bite fever in humans.

GRAM-NEGATIVE, ANAEROBIC BACTERIA: *BERGEY'S MANUAL*, PART 9

These rod-shaped or pleomorphic rods are obligate anaerobes inhabiting the intestinal tracts and other natural cavities of humans, other animals, and insects. Some species are pathogenic. They are grouped into a single family Bacteroidaceae which is subdivided on the basis of the principal end products of organisms grown in glucose and peptone-containing media. *Bacteroides fragilis* has been isolated from specimens of appendicitis, peritonitis, heart valve infections, rectal abscesses, pilonidal cysts, and wounds and lesions of the urogenital tract. *Bacteroides ruminicola*, *B. succinogenes*, and *B. amylophilis* are rumen inhabitants.

Also appended to this part of *Bergey's Manual* are a number of genera of gram-negative anaerobic rods of uncertain affiliation. Two of these genera, *Desulfovibrio* and *Selenomonas*, are fairly well known. *Desulfovibrio* spp. are important in the biogeochemical cycling of sulfur because they gain energy by anaerobic respiratory reduction of sulfates or other reducible sulfur compounds to H_2S. *Selenomonas ruminantium* is one of the more abundant species of rumen flora.

GRAM-NEGATIVE COCCI AND COCCOBACILLI (AEROBES): *BERGEY'S MANUAL*, PART 10

A single family Neisseriaceae and two agents of uncertain affiliation comprise this group of aerobic, nonmotile cocci and coccobacilli. The best-known species in the group are *Neisseria gonorrhoeae*, the organism that causes gonorrhea and other infections, and *N. meningitidis*, the agent of epidemic cerebrospinal fever. Species of *Moraxella* cause conjunctivitis in humans, cattle, horses, goats, and other animals. *Paracoccus denitrificans* and *P. halodenitrificans*, members of a genus of uncertain origin, can use either oxygen or nitrate as an electron acceptor and thus under anaerobic conditions play important roles in the nitrogen cycle.

GRAM-NEGATIVE, ANAEROBIC COCCI: *BERGEY'S MANUAL*, PART 11

These anaerobic, cytochrome oxidase negative, catalase negative, chemoorganotrophic, nonmotile diplococci belong to a very small family, the Veillonellaceae. *Veillonella parvula* is parasitic in the mouth and in the intestinal and respiratory tracts of humans and other animals.

GRAM-NEGATIVE, CHEMOLITHOTROPHIC BACTERIA: *BERGEY'S MANUAL*, PART 12

The organisms oxidizing ammonia or nitrite (family Nitrobacteraceae), metabolizing sulfur and sulfur compounds, and depositing iron and/or manganese oxides (family Siderocapsaceae) are grouped together in this part of *Bergey's Manual*.

Nitrobacteraceae. The Nitrobacteraceae are a morphologically diverse group consisting of genera of almost every shape (short rods, long rods, pleomorphic, cocci, spirillae) and flagellation (nonmotile, polar, subterminal, peritrichous). Genera are recognized on the basis of their morphology and the particular steps in the oxidation process they are able to carry out. *Nitrosomonas, Nitrosococcus, Nitrosospira,* and *Nitrosolobus,* the nitrosofying bacteria, oxidize ammonia to nitrite. *Nitrobacter, Nitrospina,* and *Nitrococcus,* the nitrifying bacteria, oxidize nitrite to nitrate. All of these organisms are chemolithotrophs, fulfilling their energy needs by the oxidation of ammonia or nitrite and their carbon needs by the fixation of carbon dioxide. Most species have extensive internal membrane systems with which the oxidase systems are associated. Nitrifying and nitrosofying bacteria are widespread in soils and water, where they play key roles in the nitrogen cycle.

Organisms Metabolizing Sulfur. A number of genera of uncertain affiliation metabolize sulfur and sulfur compounds. The best known of these is *Thiobacillus,* which derives energy from the oxidation of one or more reduced or partially reduced sulfur compounds including sulfides, elemental sulfur, thiosulfate, polythionates, and sulfite. One species, *Thiobacillus ferrooxidans,* also utilizes ferrous compounds as electron donors. The genus includes species that are strictly chemolithotrophic, some that are facultative autotrophs, and one species, *T. perometabolis,* which requires both a partially reduced sulfur compound and organic compounds for optimal growth. Most species are obligate aerobes. Various species are widely distributed in sea water, marine muds, sewage, fresh waters, soils, acid mine waters, sulfur springs, and similar habitats. *Sulfolobus acidocaldarius* is a facultative autotroph that uses elemental sulfur as an energy source. It is found in solfatara areas containing hot acid environments in both soils and water.

Organisms Depositing Iron or Manganese Oxides. The members of the Siderocapsaceae family are spherical, ellipsoidal, or rod-shaped cells that deposit iron or manganese oxides on or in their capsules. Genera differ in cell shape and in the form of the sheath. They are found in iron-rich waters and muds.

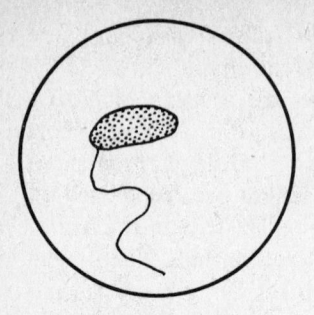

Nitrobacter winogradskyi
~ ×12,000

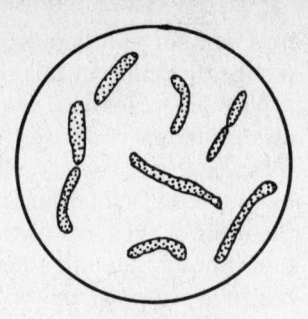

Methanobacterium thermoautrophicum
~ ×2,500

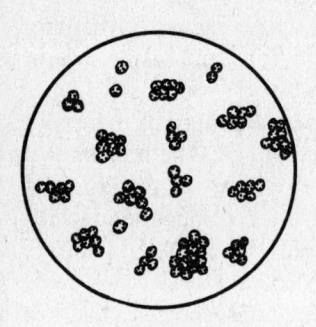

Staphylococcus aureus
~ ×1,000

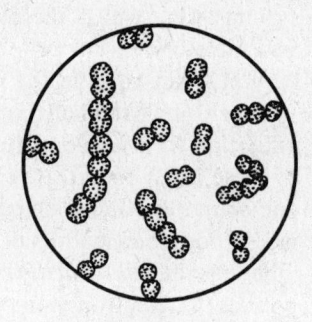

Streptococcus pyogenes
~ ×3,000

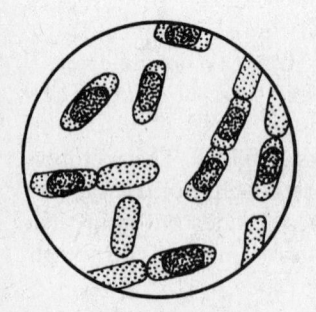

Bacillus cereus with spores
~ ×3,000

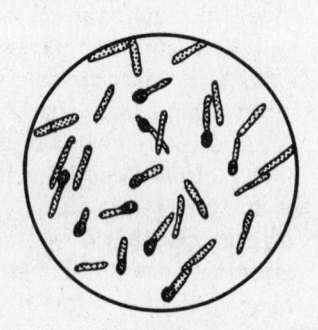

Clostridium botulinum
~ ×2,000

METHANE-PRODUCING BACTERIA: *BERGEY'S MANUAL*, PART 13

Methanogenic bacteria (family Methanobacteriaceae) in common with *Halobacterium, Sulfolobus,* and *Thermoplasma,* lack typical bacterial peptidoglycan cell walls. Methanogens have several unique coenzymes, 2-mercaptoethane-sulfonic acid (coenzyme M; involved in methyl transfer reaction) and a second yellow fluorescent heterocyclic ring system (F_{420}) that participates in electron transfer reactions, a function performed by nicotinaminde adenine dinucleotide (NAD) and flavin coenzymes in other organisms. Methanogens also have RNA sequences that are markedly different from those in other organisms.

The family includes a variety of gram-positive and gram-negative morphological types including straight rods, curved rods, and cocci occurring singly, in pairs, in clumps, and in packets. Methanogens are very strict anaerobes and obtain their energy by reducing CO_2 to CH_4 (methane) with electrons generated in the oxidation of H_2 or formate, or by the fermentation of acetate. They are unable to use carbohydrates or other organic compounds as energy sources. Many methane bacteria can grow completely on H_2 plus CO_2. About 90 percent of the CO_2 is converted to CH_4, the remainder to cell materials. Methanogens are widely distributed in nature, being found in anaerobic habitats such as sediments, soils, anaerobic sewage digestors, and gastrointestinal tracts.

GRAM-POSITIVE COCCI: *BERGEY'S MANUAL*, PART 14

The gram-positive cocci are familiar to every microbiology student. The group is subdivided into three families: Micrococcaceae, Streptococcaceae, and Peptococcaceae, the first two being aerobic and/or facultatively aerobic, the latter, aerobic.

Micrococcaeae. Micrococci are chemoorganotrophs that characteristically divide in more than one plane to form regular or irregular clusters or packets. They are catalase positive and have both fermentative and respiratory metabolic pathways. Gas is not produced in the fermentation of glucose. The family as presently conceived contains only aerobic or facultatively anaerobic forms. All *Micrococcus* species are aerobic. They are common inhabitants of soils, fresh water, and on the skin of humans and other animals. None are pathogenic. All micrococci can grow in the presence of 5% NaCl and many can tolerate 10 to 15% salt concentrations. *Staphylococcus* spp. are facultative anaerobes and are mainly associated with skin, skin glands, and mucous membranes of warm-blooded animals. Many strains are potential pathogens. As mentioned earlier in this book (Chapter 11), *Staphylococcus aureus* strains under suitable conditions produce a variety of enzymes and toxins that play roles in initiating and furthering their pathogenicity. In addition, strains of *S. aureus* produce enterotoxins that cause food poisoning.

Streptococcaceae. Members of this family divide in one or two division planes forming pairs, chains of varying lengths, or tetrads. They are all facultatively anaerobic, fermentative chemoorganotrophs forming lactic, acetic, and formic acids and ethanol and CO_2 from carbohydrates. The family is subdivided into genera on the basis of cell division and fermentative patterns. *Streptococcus* spp. ferment glucose by the Embden-Meyerhof pathway and divide in one plane, producing pairs and chains. Many *Streptococcus* spp. (e.g., *S. pyogenes, S. equisimilis, S. zooepidemicus, S. equi, S. sanguis, S. pneumoniae*) have strains that produce a variety of enzymes and toxins which play roles in initiating and furthering their pathogenicity in humans and other animals. Other species and strains live as harmless saprophytes on and in the upper respiratory, digestive, and urinogenital tracts of humans and other animals or are associated with milk and fermentation of milk products (e.g., *S. thermophilus, S. diacetilactis, S. lactis, S. cremoris*). *Leuconostoc citrovorum* and *L. dextranicum* are used as butter starter cultures. *L. mesenteroides* is introduced in the early and intermediate stages of the fermentation of sauerkraut, pickles, and olives. *Pediococcus cerevisiae* is an organism used in starter cultures for the preparation of fermented sausages. It is also found in spoiled beer and brewers yeast preparations.

Peptococcaceae. These bacteria occur in pairs, tetrads, and irregular masses. The various species of the genera *Peptococcus* and *Peptostreptococcus* are found in the mouth, intestinal, upper respiratory, and urogenital tracts of humans and other animals where they are sometimes associated with pathological conditions. *Ruminococcus* spp., whose habitat is in the rumen, are strictly anaerobic. They actively digest cellulose and ferment cellobiose.

ENDOSPORE-FORMING RODS AND COCCI: *BERGEY'S MANUAL*, PART 15

The endospore-forming bacteria are placed in a single family Bacillaceae which is subdivided by the shape of the cells and oxygen requirements. *Bacillus* species are rod-shaped aerobic or facultative anaerobes. Most species produce catalase and are widely distributed in soils, dust, and aquatic habitats. Various strains produce polypeptide antibiotics, the best known of which is polymyxin produced by *B. polymyxa*. Many species grow in food and spoil it. *B. anthracis* causes anthrax in humans and other animals. *B. thuringiensis* is pathogenic for the larvae of lepidoptera, a quality that is responsible for its employment as a microbial insecticide. Several other species, *B. popilliae* and *B. lentimorbus,* cause disease in beetles and their grubs and are potential insecticides against the Japanese beetle. Another species, *B. larvae,* causes foulbrood, a lethal disease of honeybee larvae. *Clostridium* spp. are strictly anaerobic and do not reduce sulfate. They are common in soil, marine and freshwater sediments, and in the intestinal tract of humans and other animals. They have also been isolated from food. Many species fix nitrogen. *Clostridium botulinum* forms the well-known botulism toxin in improperly canned foods. *C. perfringens, C. histolyticum, C. novyi,* and other species have been isolated from infected gangrenous wounds of humans and other animals. *C. tetani* causes tetanus in wound infections. *C. cellobioparum,* a species that ferments cellulose, is prominent in the rumen flora.

GRAM-POSITIVE, ASPOROGENOUS, ROD-SHAPED BACTERIA: *BERGEY'S MANUAL*, PART 16

The family Lactobacillaceae contains only three genera of uncertain affiliation, *Listeria, Erysipelothrix,* and C*aryophanon*. Lactobacilli are straight or curved nonmotile rods that occur singly or in chains. They are anaerobic or facultative anaerobes with complex organic nutritional requirements. Most species are highly saccharoclastic and produce large quantities of lactate as an end product. Lactobacilli are found in dairy products, grain and meat products, beer, wine, fruits, fruit juices, and vegetables; in water and sewage; and in the mouth, intestinal tract, and vagina of humans and other animals. Many species are employed in the processing of food and industrial processes. *Lactobacillus bulgaricus* is used in the manufacture of yogurt, cheese, and similar milk products. *L. lactis* and *L. helveticus* are also involved in cheese manufacture. *L. plantarum* is involved in the final stages of sauerkraut, pickle, and olive fermentation. *L. delbrueckii* has been used in industral fermentations yielding lactic acid. *Listeria monocytogenes* causes meningitis, encephalitis, septicemia, endocarditis, abortion, and abscesses and local purulent lesions in humans and many other animals. *Erysipelothrix rhusiopathiae* is a widely spread pathogen of mammals, birds, and fish. Fish handlers occasionally get erysipeloid infections derived from fish.

ACTINOMYCETES AND RELATED ORGANISMS: BERGEY'S MANUAL, PART 17

This is a very large group of microorganisms which includes the coryneform group, the family Propionibacteriaceae, and the order Actinomycetales which contains eight families: Actinomycetaceae, Mycobacteriaceae, Frankiaceae, Actinoplanaceae, Dermatophilaceae, Norcardiaceae, Streptomycetaceae, and Micromonosporaceae.

The Coryneform Group. The coryneform group contains gram-positive, acid-fast negative, rod-shaped organisms, often with granules and irregularly staining segments. The cornyebacteria are divided into three groups: (1) pathogens or parasites of humans and other animals, (2) plant pathogens, and (3) nonpathogenic forms. The cells of human and animal pathogens and parasites frequently have club-shaped swellings, and their "snapping" mode of division produces angular and palisade (picket fence) arrangements of cells. They are aerobes or facultative anaerobes. The best known of this group of human and animal parasites is *Cornyebacterium diphtheriae*. This species has been isolated from cases of diphtheria and from the nasopharynx of healthy carriers. Many other species (e.g., *C. xerosis, C. pseudodiphtheriticum*) are harmless inhabitants of the skin and mucous membranes of humans and other animals. Still other species (e.g., *C. renale, C. bovis, C. equi*) have been isolated from cases of cystitis, pyelonephritis, and mastitis in cattle; bronchopneumonia and abortion in horses; ulcerative lymphangitis, abscesses, and other chronic purulent infections in sheep, goats, horses, and other warm-blooded animals. The plant pathogens are

Actinomycetes and Related Organisms

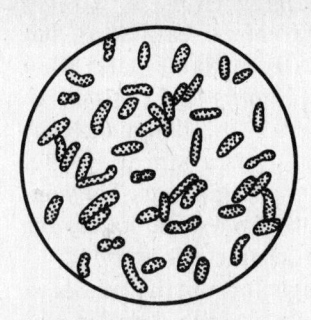

Lactobacillus acidophilus
~ X 2,500

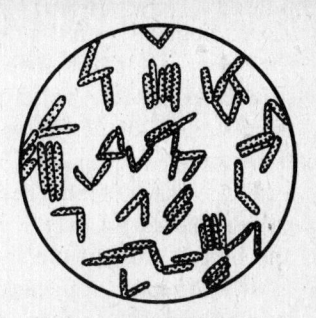

Corynebacterium diphtheriae
~ X 2,000

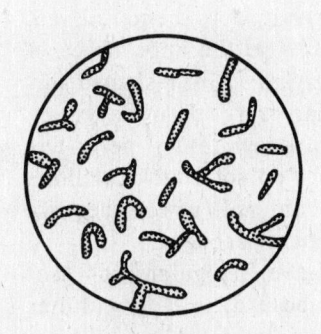

Actinomyces israelii
~ X 2,500

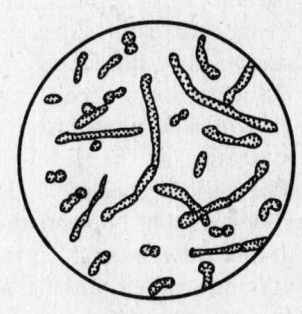

Rothia dentocariosa
~ X 2,500

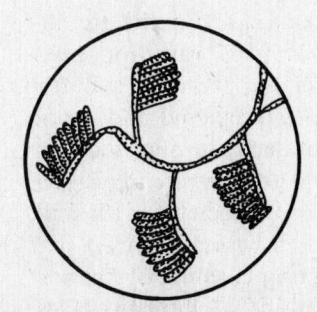

Planomonospora parontospora
~ X 600

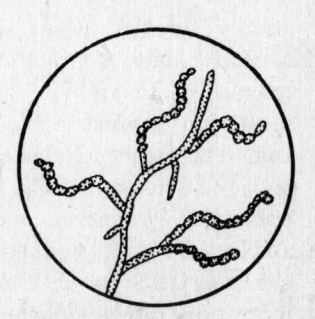

Streptomyces
~ X 2,500

morphologically similar to the human and animal pathogens, but they are less pleomorphic (morphologically variable) and have smaller palisades. Various species (e.g., *C. rathayi, C. tritici, C. iranicum, C. betae, C. oortii*) cause diseases (e.g., gumming, wilting, rotting, canker, leaf and fruit spot) in grasses, wheat, potato, alfalfa, tomato, beans, beets, and various flowering plants. The nonpathogenic *corynebacteria* have been isolated from soils, water, air, and blood. Most of the species are not well characterized.

Arthrobacter spp. have cells that characteristically undergo marked changes in form during their growth cycle in complex media. Older cultures (2 to 7 days) are composed entirely or largely of coccoid cells. When the cells are transferred to fresh medium they swell and then elongate in one, two, or more parts in the cell. Subsequent growth and division give rise to irregular rods that vary considerably in size and shape from straight to bent, curved, wedge-, or club-shaped forms. Some cells give the appearance of rudimentary branching. Many cells are arranged in V-shaped formations. Some species have a single polar or subpolar flagellum. They are not acid fast. Their metabolism is respiratory, never fermentative, and they are strict aerobes. They do not attack cellulose. The various species have been isolated from soil, where they are very common. None of them is known to be a pathogen.

Cellulomonas spp. are also gram-positive, irregular rods that may be straight, angular, club-shaped, or beaded. In older cultures the rods become shorter and a small proportion of coccoid cells are formed. These organisms, which are found in soils, attack cellulose.

Propionibacteriaceae. Bacteria in this family are gram-positive, nonsporeforming, and pleomorphic; they may be branching, filamentous, or rod-shaped. In liquid cultures the cells occur in "V" and "Y" configurations, short chains, or clumps, giving the appearance of Chinese characters. *Propionibacterium* spp. vary from aerotolerant to anaerobic. They are chemoorganotrophs and their fermentation products include combinations of propionic and acetic acids and frequently lesser amounts of isovaleric, formic, succinic, or lactic acids and CO_2. *Propionibacterium freudenreichii, P. globosum, P. shermanii,* and other species are associated with milk and certain milk products. The flavor and the holes ("eyes") that characterize Swiss cheese are produced during ripening by the fermentations of these bacteria; accumulations of CO_2 form the holes. Other species (e.g., *P. avidum, P. acnes,* and *P. lymphophilum*), some of which may be pathogenic, have been isolated from the skin of humans and the intestinal tract of humans and animals where they may be associated with infected wounds, tissue abscesses, pus, and urinary tract infections.

Eubacterium spp. are obligately anaerobic, nonsporeforming, gram-positive, pleomorphic rods. They are chemoorganotrophs and differ from other gram-positive nonsporeforming rods in that they produce large amounts of butyric, acetic, or formic acids. Other genera produce large amounts of propionic acid (*Propionibacterium*); only lactic acid (*Lactobacillus*); succinic (in the presence of CO_2) and lactic acids with small amounts of acetic or formic acids (*Actinomyces*); or acetic and lactic acids with or without formate as sole major acid products (*Bifidobacterium*). *Eubacterium* spp. are found in cavities of humans and other animals, plant and animal products, infections of soft tissue, and in the soil. Some species may be pathogenic. For example: *Eubacterium ruminantium* is common in the bovine rumen; *E. cellulosolvens* is found in the sheep rumen; *E. alactolyticum* has been found in dental tartar and in various types of absceses of the brain, lung, intestinal tract, and mouth; and *E. helminthoides* has been isolated from the mouth of nursing infants, the intestines of mollusks, and pond mud.

Actinomycetales. The order Actinomycetales contains eight families of gram-positive bacteria that tend to form branching filaments, which in most families develop into a mycelium. In some families, spores are formed on aerial and/or substrate hyphae. Except for genera in the family Actinomycetaceae, all members of the order are aerobic. Actinomycetes are common in soils; they are less frequently isolated from fresh water and pathogenic conditions in plants and animals. The order is divided into families on the basis of the nature of the mycelium, if it is formed, the types of sporulation, and other propagules (propagating units) that may be formed.

ACTINOMYCETACEAE. Bacteria belonging to the family Actinomycetaceae do not form mycelia, but they do form branched filaments at some stage. The filaments tend to fragment into diphtheroid or coccoid forms. None of the members of the family are acid fast or motile. Most are facultative anaerobes but some are anaerobes or aerobes. They are chemoorganotrophs that ferment carbohydrates; end products include acetic, formic, lactic, and succinic acids, but not propionic acid. Proteolytic activity is rare. Perhaps the best-known species in the family is *Actinomyces bovis* which causes a disease, actinomycosis ("lumpy jaw") in cattle, swine, horses, other animals, and occasionally in humans. The initial lesion of *A. bovis, A. israelii,* or *Arachnia propionica* involves the face, neck, tongue, or mandible. Involvement of lungs is frequent. *A. odontolyticus, A. naeslundi,* and *A. viscosus* have been isolated from periodontal disease, dental plaque, and deep dental caries. *Rothia dentocariosa* is a common inhabitant of the normal mouth and throat that morphologically resembles *Actinomyces* spp. but grows better aerobically and differs in physiological and cell wall characteristics.

MYCOBACTERIACEAE. The family Mycobacteriaceae contains a single genus *Mycobacterium.* Mycobacteria are slightly curved or straight rods which sometimes branch. Mycelium-like growth may occur, but slight disturbance usually causes it to fragment into rods or coccoid elements. *Mycobacterium* are acid fast at some growth stage. They do not form endospores, conidia, capsules, or aerial hyphae. The lipid content of cells and cell walls is high. Several species of mycobacteria are very well known because of the diseases they cause. *M. tuberculosis* (and several other species of *Mycobacterium*) produces tuberculosis in humans, other primates, dogs, and some other animals that have contact with humans. *M. bovis* produces tuberculosis in cattle, both domestic and wild ruminants, humans and other primates, carnivores including dogs and cats, swine, parrots and some other birds, rabbits, and other animals. *M. marinum,* a species isolated from diseased fish, frequently causes skin lesions on the elbow, knee, foot, finger, or toe where an abrasion has occurred in a swimming pool harboring the organism. *M. avium* causes tuberculosis in birds but rarely in cattle, swine, and other animals. *M. leprae* causes leprosy in humans.

FRANKIACEAE. The Frankiaceae, a family with only one genus, *Frankia,* has species that are symbiotic, filamentous, myceliumforming bacteria which induce and live in root nodules on a wide variety of nonleguminous dicotyledonous plants. The nodules fix molecular nitrogen. These bacteria also have a free stage in the soil. The various species of *Frankia* are separated from each other by the types of host plants with which they become associated.

ACTINOPLANACEAE. The family Actinoplanaceae contains 10 genera that are abundant in all types of soils containing humus and plant litter, and less common in freshwater habitats and in animal material such as hair and other keratin and chitin substrates. Members of the family form characteristic aerial hyphae and produce spores arranged in the sporangia in one or more spirals or in parallel rows. Some genera produce polarly flagellated motile spores (zoospores). The various genera are separated from each other on the basis of sporangia and spore characteristics. Some species (e.g., *Kitasatoa purpurea, Planomonospora parontospora*) produce antibiotics (e.g., chloramphenicol, sporangiomycin).

NOCARDIACEAE. The two genera of the family Nocardiaceae are also separated from each other by mycelial and sporangial characteristics. Their cell walls are distinguished by components of *meso*-diaminopimelic acid, arabinose, and galactose. In the tropics, two species, *Nocardia asteroides* and *N. brasiliensis,* are pathogens producing nocardiosis, a usually opportunistic pulmonary disease that may spread to other parts of the body, particularly the brain and kidney. Many isolates are acid fast. Most species of *Nocardia* have been isolated from soils. An antibiotic, Neonocardin, is produced by *N. kuroishii.*

STREPTOMYCETACEAE. Organisms in the family Streptomycetaceae are naturally abundant in soils and produce a well-developed, branched mycelium that does not readily fragment. Aerial spores are produced. Interest in this family is great because many strains of *Streptomyces* species and other genera in the family produce one or more antibacterial, antifungal, antialgal, antiviral, antiprotozoal, or antitumor antibiotics. In one study, close to 50% of the strains of *Streptomyces* isolated from soils produced antibiotics.

MICROMONOSPORACEAE. Micromonosporaceae is another family of aerobic, primarily saprophytic soil forms. The six genera in this family are distinguished on the basis of mycelium, sporangium, and cell wall characteristics. Some strains of species in this family (e.g., *Micromonospora purpurea*) also produce antibiotics (e.g., gentamycin complex).

RICKETTSIAS: *BERGEY'S MANUAL,* PART 18

Rickettsiales. True rickettsias are treated as members of a single order, Rickettsiales. The majority are rod-shaped, coccoid, and often pleomorphic, nonmotile microorganisms that multiply only inside host cells. They have cell walls and are gram-negative. Many species have been cultivated in embryonated hen's eggs and/or in vertebrate tissue culture. Rickettsias are intracellular parasites that divide by binary fission in the reticuloendothelial tissue of mammals and other vertebrates. They are transmitted naturally by blood-sucking arthropods (fleas, lice, ticks). Unlike viruses, rickettsias have metabolic enzymes including those of the Krebs (TCA) cycle, electron transport, and many enzymes required for biosynthesis of macromolecules. Experimental studies have shown that the living host must supply ATP, NAD, CoA, and other required metabolic components; these pass (leak) through the rickettsial cell wall and membrane. It is speculated that rickettsias evolved from other bacterial intracellular parasites by the progressive loss of metabolic capabilities (degenerative evolution).

The order is divided into three families; Rickettsiaceae, Bartonellaceae, and Anaplasmataceae.

RICKETTSIACEAE. These bacteria are associated with tissue cells other than erythrocytes or with certain organs in arthropods. They are rarely extracellular in arthropods. *Rickettsia typhi* causes murine (endemic) typhus. Rats and other rodents are the primary reservoirs of the organism. It is transmitted mainly by the rat louse *Polyplax spinulosus* and the rat flea *Xenopsylla cheopis;* the human flea *Pulex irritans* and the human louse *Pediculus humanus* also play roles in natural transmission to humans. *R. rickettsii* causes Rocky Mountain spotted fever. It is transmitted by various ticks (e.g., the wood tick *Dermacentor andersoni,* the dog tick *D. variabilis,* the lone-star tick *Amblyomma americanum,* and the rabbit ticks *Haemaphysales leporispalustris* and *D. parumapertus*). *R. tsutsugamushi* is the agent of scrub typhus. It is transmitted by various mites, or chiggers (e.g., *Leptotrombidium akamushi, L. deliense, L. pallidum, L. scutellare*). Trench fever, a disease transmitted by the body louse, is caused by *A. quintana. Coxiella burnetii* causes Q fever, a moderately severe but rarely fatal pulmonary infection of humans. It is transmitted by aerosols. *Ehrlichia canis,* an agent transmitted by dog ticks (*Rhipicephalus sanguineus*), causes canine ehrlichiosis. *E. phagocytophilia* infects sheep and cattle. *Symbiotes lectularius* are rickettsia-like pleomorphic intracellular symbionts in mycetomes and other body tissues of bedbugs (*Cimes*). *Blattabacterium cuenoti* is an intracellular symbiont in mycetocytes in abdominal fat body tissue and in ovaries and eggs of cockroaches (*Blatta*). *Rickettsiella popilliae* is a parasite of the Japanese beetle (*Popillia*).

BARTONELLACEAE. Members of the family Bartonellaceae are

parasites of human erythrocytes and those of other vertebrates. They have gram-negative cell walls, the formation of which can be inhibited by penicillin. Their shape varies from rods, cocci, rings, disks, beaded forms, and filaments. The erythrocytic forms are visualized in Giemsa preparations, a common procedure used to differentiate blood cells and protozoan parasites. *Bartonella bacilliformis* causes a progressive anemia (oroya fever) and a cutaneous eruption (verruca peruana) which usually follow each other. The organisms, which are often flagellated, are isolated from clinical specimens of blood and endothelial cells of lymph nodes, spleen, and liver. *B. bacilliformis* is transmitted by the bite of sand flies (*Phlebitinys verrucarum*). It has been cultivated *in vitro* on nonliving media. *Grahamella* spp. parasitize the erythrocytes of rodents and other animals.

ANAPLASMATACEAE. In contrast with the two other families, members of the Anaplasmataceae have not yet been cultivated. They are obligate parasites that are found within or on the erythrocytes or plasma of various wild and domestic vertebrates. In Giemsa-stained blood smears, they appear as rod-shaped, spherical, coccoid-, or ring-shaped bodies. Their internal structure is similar to that of other rickettsias. *Anaplasma marginale* causes severe bovine anaplasmosis in cattle, buffalo, bison, antelopes, deer, camels, and other ruminants of tropical and subtropical regions. *A. ovis* causes a similar disease in sheep, deer, and goats. These organisms are transmitted by ticks and other biting arthropods. Ticks also spread *Aegyptianella pullorum,* the cause of aegyptianellosis in species of wild and domestic birds (e.g., geese, ducks, turkeys, guinea fowls, pigeons, quail).

Chlamydiales. The order Chlamydiales consists of but a single genus with two species. These are very small (0.2 to 1.5 μm) coccoid microorganisms whose obligately intracellular life cycle has two phases, a small rigid-walled infectious form (elementary body) which changes into a larger, thin-walled, noninfectious form (initial body) that divides by fission. The life cycle is completed when daughter cells reorganize to again form elementary bodies. *Chlamydia trachomatis* causes a variety of oculo-urogenital diseases: trachoma, inclusion conjunctivitis, lymphogranuloma venereum, urethritis, and proctitis. Occasionally it is associated with arthritis. Transmission of the disease from person to person occurs by contact contamination of the conjunctiva, the oral or genital mucous mem-

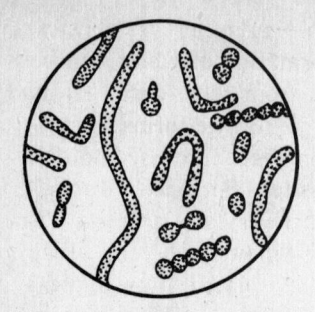

*Mycoplasma
mycoides*
~ X 9,000

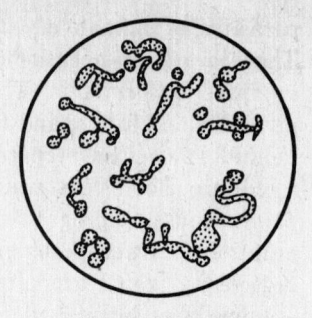

*Mycoplasma
pneumoniae*
~ X 5,000

branes during birth, or during sexual and other contacts. *C. psittaci* parasitize the tissue cells of more than 100 species of wild and domestic birds and mammals where they cause psittacosis ornithosis; pneumonitis in cattle, sheep, goats, horses, and pigs; placentitis leading to abortion in cattle and sheep; enteritis in cattle and hares; conjunctivitis in guinea pigs, cattle, and sheep; encephalitis in opossums; encephalomyelitis in cattle; and polyarthritis in sheep, cattle, and pigs. Transmission is accomplished by inhalation of dust containing bird excrement (psittacosis); transovarian passage, veneral contact, arthropod bites, and fecal-oral transmission cycles.

MYCOPLASMAS: *BERGEY'S MANUAL*, PART 19

Procaryotes bound only by a unit membrane and walls at any stage in their life cycles are placed in the class Mollicutes. They are small (sometimes ultramicroscopic; 100 to 250 nm) and highly pleomorphic, varying in a single culture from tiny coccoid elements to branching filaments. The small coccoid elements are the smallest known cells capable of independent growth. The single class Mollicutes has a single order Mycoplasmatales with two families, Mycoplasmataceae and Acholeplasmataceae, which are distinguished from each other by the fact that Mycoplasmataceae require sterol (cholesterol) for growth. This is in itself unusual since no other procaryote group seems to require sterols for its cell membranes. *Mycoplasma mycoides* causes contagious pleuropneumonia in cattle. Other species of *Mycoplasma* cause pneumonia, conjunctivitis, arthritis, vaginitis, or encephalitis in mice, rats, sheep, cattle, goats, pigs, turkeys, chickens, and cats. Species of *Acholeplasma* live as saphrophytes as well as mammalian and avian parasites. *Spiroplasma citri* and other "mycoplasma-like bodies" have been isolated from various types of plant diseases (e.g., "stubborn disease" of citrus, aster yellows, and corn stunt).

13

EUCARYOTIC MICROORGANISMS

Eucaryotic organisms represent a diverse spectrum of cellular types. Along with procaryotes they are found in almost every conceivable habitat ranging from the snows and barren rocks on the world's highest mountains to the deepest seas, from Arctic ice to desert sands, and on the surfaces and in the tissues and body cavities of all kinds of multicellular animals and plants. As indicated in Chapter 1, many groups of eucaryotic organisms do not fall into the "classical" three major categories: algae, fungi, and protozoa. Drawing lines between many chloroplast-bearing species and closely related nonpigmented forms to separate them into algae or protozoa seems quite artificial and unnecessary to many microbiologists. Both protozoologists and mycologists (scientists who study molds) lay claim to a number of amoeboid groups. In the final analysis, microbiologists must accept the arbitrariness of the taxonomic system and devote their efforts to more fruitful enterprises.

ALGAE

Algae are chlorophyll-bearing organisms that lack true roots, stems, leaves, and tissues. Though most forms are microscopic and unicellular, others are multicellular and some of the giant kelps (seaweed) exceed 30 m in length. The larger algae (macroalgae, macrophytes) are not considered to be microorganisms by any stretch of the definition. They will be treated briefly so that the smaller algae can be considered in context with the entire group.

Reproduction is asexual and/or sexual. Asexual reproduction may be by vegetative propagation or by means of unicellular spores. These spores are of various types: *zoospores,* which are motile and

Algae

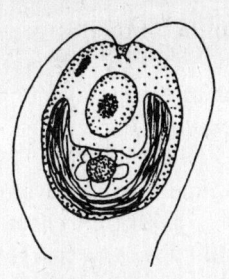

Chlamydomonas
∼ ✕ 1,500

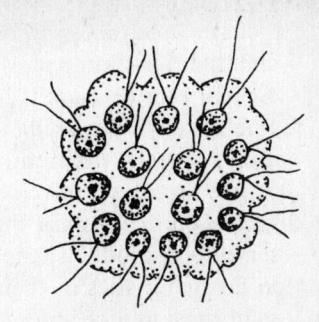

Gonium pectorale
∼ ✕ 475

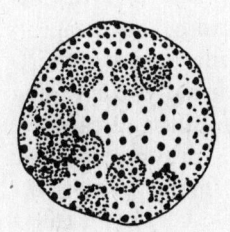

Volvox aureus
with autocolonies
∼ ✕ 200

Chlorella
∼ ✕ 2,000

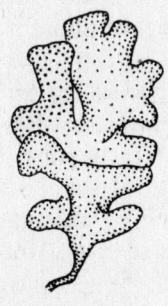

Ulva lactuca
¼ life size

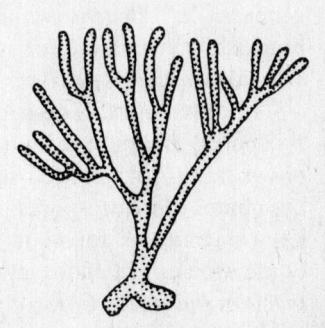

Codium
1/10 life size

flagellate, and *aplanospores,* which are nonmotile. *Akinetes* are nonmotile, resistant spores.

Sexual reproduction is by the union of gametes to form a zygote. Gametes may be *isogamous* (not sexually differentiated) or *anisogamous* (sexually differentiated).

Algae are a very diverse group. They are classified into phyla or divisions on the basis of their primary and accessory pigments; their reserve food products; the number, morphological type, and insertion characteristics of their flagellae; chloroplast fine structure; and the physical and chemical characteristics of their cell walls. Higher levels of algal taxonomy have been in considerable flux since the electron microscope came into wide general use in the study of their fine structure.

Chlorophyta (Chlorophycophyta): Green Algae. (Fig. 13.1) Chlorophyta or green algae have both chlorophylls a and b; they generally have two identical, equally long, smooth flagella apically inserted, and a double- or triple-layered cell wall usually containing cellulose (some green algal groups have mostly pectin and callose). Food is stored as starch. Some of the green algae form macroscopic filaments. The green algae are separated into 15 major groups (orders) but only 7 of them, Volvocales, Chlorellales, Ulvales, Caulerpales, Siphonocladales, Dasycladales, and Zygnematales, contain species that are widely known among nonspecialists.

VOLVOCALES. This order includes free-swimming flagellated cells (*Carteria, Chlamydomonas*), simple colonies (*Gonium, Pandorina, Pascherina*), and large colonies with thousands of cells, cellular specialization, and anisogamous reproduction (*Eudorina, Pleodorina, Volvox*). Most volvocalean algae are found in fresh water but many genera (i.e., *Brachiomonas, Dunaliella, Chlamydomonas*) have brackish, hypersaline, or marine species. Protozoologists also consider this order as part of their group.

CHLORELLALES. These chlorophytes lack the capacity to form zoospores, and only a few (e.g., *Golenkinia, Scenedesmus, Eremosphaera*) form flagellated gametes. In most species, reproduction is by nonmotile autospores and autocolonies. *Chlorella,* the best-known genus, is found in soil, fresh water, and marine habitats. Some species are endosymbiotic within invertebrates such as *Paramecium bursaria, Chlorohydra viridis,* and sponges.

ULVALES (ULOTRICHALES). Ulvales are commonly known as sea lettuce. A few freshwater species are also known. They form expanded sheets, narrow ribbons, and hollow tubes which are attached to rocks, dead shells, and wharf pilings. Many species propagate by vegetative fragmentation. Biflagellated zoospores and gametes are also formed. In some genera there is an alternation of similar appearing (isomorphic) generations in which the diploid generation develops from the zygote and the haploid generation develops from a zoospore. Species of *Ulothrix* are common in freshwater habitats. They are often attached to rocks in streams.

CAULERPALES, SIPHONOCLADALES, AND DASYCLADALES. These are macroscopic green seaweeds with a siphonous or coenocytic (multinucleate without cross cell walls) construction. These plants

Life Cycles of 2 Green Algae

a) Chlamydomonas

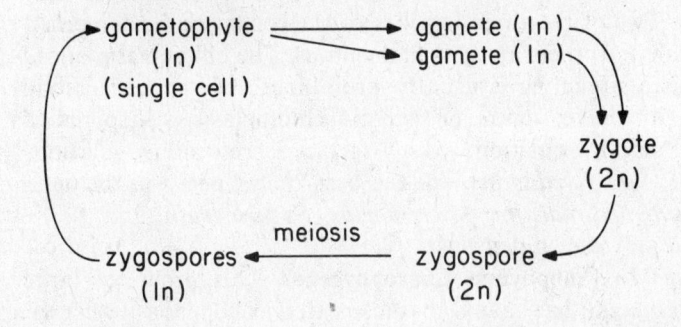

gametophyte (1n) (single cell) → gamete (1n), gamete (1n) → zygote (2n) → zygospore (2n) → meiosis → zygospores (1n) → gametophyte

b) Cladophora

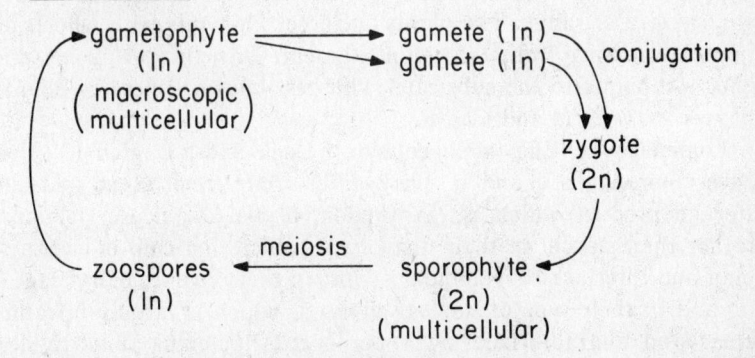

gametophyte (1n) (macroscopic multicellular) → gamete (1n), gamete (1n) — conjugation → zygote (2n) → sporophyte (2n) (multicellular) → meiosis → zoospores (1n) → gametophyte

Fig. 13.1. Life cycles of two green algae.

have tubular filaments that lack cross walls (except for reproductive structures). The three orders are distinguished from each other by their patterns of cell division and thallus organization. *Codium* is a widespread spongelike marine genus which in some regions (e.g., Atlantic coastline) is considered a nuisance seaweed. *Halimeda* and *Penicillus* ("Neptune's shaving brush") are marine calcareous species commonly encountered by snorkelers and scuba divers in shallow tropical and subtropical waters. *Acetabularia* "mermaid's wine glass") is a distinctive umbrellalike alga which has been the focus of many morphogenetic studies.

ZYGNEMATALES. These forms are exclusively freshwater; when abundant they form slimy pond scums. Conjugation, a characteristic of the order, is accomplished by vegetative cells moving toward one another by amoeboid movement. Filamentous species may reproduce asexually by fragmentation. Under appropriate environmental conditions, the cell walls of vegetative cells thicken to form *akinetes*. Two of the three families of the order are fundamentally unicellular; the other family is filamentous. The chloroplasts of zygnematalean algae have usually prominent, bar-shaped, broad-shaped, ribbonlike, spiral, or asteroid chloroplasts with pyrenoids (areas of starch formation). Many species secrete slimy, gelatinous materials. *Spirogyra* is perhaps the best-known genus in the order. *Closterium, Cosmarium, Micrasterias, Staurastrum,* and *Desmidium* are well-known desmids.

Charophyta (Subphylum Charophyceae). This group of larger freshwater algae has many characteristics in common with green algae (e.g., chlorophylls a and b, starch storage) and some unique characteristics of its own. None of these algae are in the microorganism size or range. The plants are erect filaments with whorls of branches arising from the main filament. Branching rhizoids anchor the plants to the substrate. The best-known genera of charophytes are *Nitella* and *Chara*.

Euglenophyta (Euglenophycophyta): Euglenoids. Euglenoids also have chlorophylls a and b. They differ from green algae in their storage product which is, paramylon, a β-1, 3-glucose polymer, rather than starch; in their distinctive pellicle built up of overlapping and interlocking pellicular strips; in their coarse, hairy flagella; and in their type of nuclear division, which is largely intranuclear and characterized by a persistent nucleolus that divides during mitosis and by the arrangement of the chromosomes (long

Algae

Acetabularia cap
~ X 2

Spirogyra
~ X 200

Closterium
~ X 300

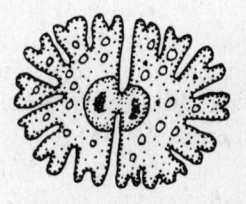

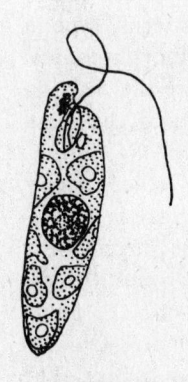

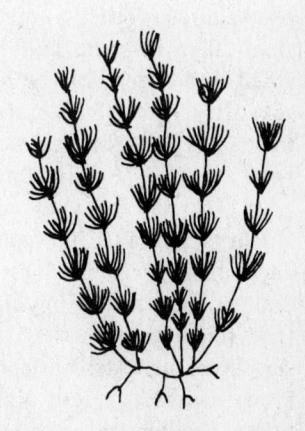

Micrasterias
~ X 200

Euglena
~ X 500

Nitella
⅓ life size

axis parallel to the long axis of the microtubules in the division figure). Flagella arise anteriorly. When there are two, one or both may arise from an invagination which consists of a canal and a reservoir. The contractile vacuole discharges into the reservoir. The flagellum is thickened (flagellar swelling) in the reservoir. Many euglenoids undergo a peculiar type of stretching, bulging, and expanding, known as euglenoid movement (metaboly). Euglenoids are widely distributed in fresh water and marine habitats and on moist soils and muds. There are many colorless forms. Protozoologists consider euglenoids to be part of the Sarcomastigophora. Among the best-known genera are *Euglena, Astasia, Peranema,* and *Trachelomonas.*

Phaeophyta (Phaeophycophyta): Brown Algae. Phaeophytes, or brown algae, are familiar seaweeds on rocky shores from the upper littoral zone to the sublittoral zone. In clear tropical waters, they are distributed throughout the entire euphotic zone down to the depths of 100 m. A few genera are freshwater. Brown algae are characterized by a distinctive group of photosynthetic pigments: chlorophylls a and c, β-carotene; violaxanthin; and fucoxanthin. Their food reserves, which can be considerable (up to 35% of their dry weight), are laminarin (β-1,3-glucan with some β-1,6 linkages). A large amount ($\sim$ 25%) of alginic acid, a polymer of D-mannuronic and L-glucuronic acids, is found in cells walls and intracellular spaces of many large species. Brown algae vary in size from microscopic epiphytes (organisms that live on the surfaces of larger plants or seaweeds) to giant kelp (*Macrocystis*), which reach more than 50 m in some "underwater forests." Brown algae are subdivided into orders on the basis of their life history patterns, type of sexuality, growth pattern, pyrenoids, and other morphological traits. Perhaps the best known of the brown algae are the *Sargassum* "weeds," *Fucus* ("rock weed"), *Ascophyllum,* and the giant kelps *Macrocystis, Nereocystis,* and *Laminaria.*

Chrysophyta (Chrysophycophyta, Heterokonts): Golden Algae. This major group of yellow-green and golden-brown algae is considered a kingdom, a phylum, or a division by various phycologists (the scientists who study algae). Three major subgroups (phyla, subphyla, or classes depending upon classification system being used) are recognized: Xanthophyta (yellow-green algae), Chrysophyta (golden-brown algae), and Bacillariophyta (diatoms). The Phaeophyta (brown algae) are considered part of the group by some experts. The Haptophyta, Eustigmatophyta, and Raphidophyta are also sometimes included. Almost the entire group is microscopic. With the exception of the diatoms and nonmotile forms in other groups, the protozoologists also lay claim to these organisms.

Several characteristics unite this large assemblage. They have similar, if not identical, pigmentation including chlorophylls a and c (some have e), diadinoxanthin, diatoxanthin, β-carotene, lutein, and fucoxanthin. The golden-brown or yellow hue of the algae is due to the presence of large amounts of fucoxanthin and other carotenoids in relation to chlorophylls. Food reserves are stored as leucosin. Another common characteristic shared by many of these organisms is that they have siliceous cell walls, scales, bristles, and other elaborations on their cell surfaces. Asexual reproduction is common; aplanospores, autospores, and zoospores are formed in various groups. Fragmentation of filaments is another asexual reproductive method.

XANTHOPHYTA (XANTHOPHYCOPHYTA; XANTHOPHYCEAE): YELLOW-GREEN ALGAE. These are largely a freshwater group. The motile cells bear two different types of anteriorly attached flagella: one is smooth (*acronematic*) and the other (*pleuronematic*) bears hairlike appendages called *mastigonemes*. The flagella are usually unequal in length and beat independently of each other (*heterodynamic*). None of the xanthophyta are well known to the general biological community. *Chlorobotrys, Tribonema, Botrydium,* and *Vaucheria* are perhaps the best known to algal specialists.

EUSTIGMATOPHYTA. Eustigmatophytes have an eyespot independent of the chloroplast, consisting of an irregular group of globules located at the anterior end of the cell adjacent to the flagellum. The pleuronematic flagellum has a distinctive swelling at its base. Genera assigned to this small class (phylum) include *Pleurochloris, Ellipsoidion,* and *Vischeria*. None of the group is really well known to nonspecialists.

CHRYSOPHYTA (CHRYSOPHYCEAE): GOLDEN-BROWN ALGAE. This group is considered a phylum, subphylum, division, or class, depending upon the taxonomic scheme used. The motile forms are also claimed by protozoologists. Chrysophytes are widely distributed in both fresh water and marine habitats. Two widely accepted schemes are used in classifying the subdivisions of chrysophytes. The older scheme includes organisms that produce motile cells with

Peranema
∼ X 450

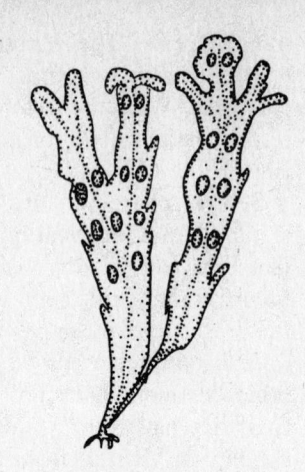

Fucus
⅓ life size

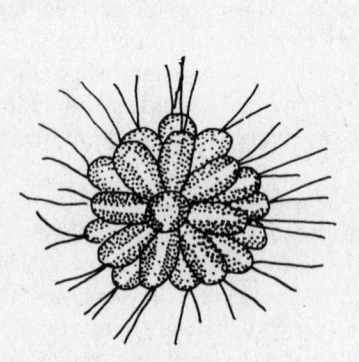

Laminaria
∼ X 1/100
life size

Synura
∼ X 600

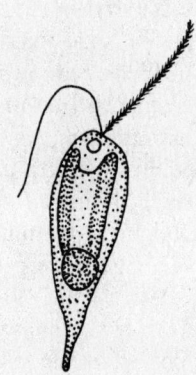

Ochromonas
∼ X 2,500

two acronematic flagella and a coiled filamentous structure known as a *haptonema*. Newer schemes consider this group as a separate phylum, Haptophyta (or class, Haptophyceae), also known as Prymnesiophyceae. The various orders of chrysophytes are separated from each other by the absence or presence of flagella, the number of flagella, and the morphology of the dominant phase in the life cycle (monad, coccoid, filamentous, thalloid, palmelloid, and colonial forms).

Ochromonadales. This group consists primarily of freshwater forms with two flagella that are heterodynamic and unequal in length. *Ochromonas* has been used in many biochemical and cell physiological studies as a model "animal" cell. *Mallomonas* cells are covered with siliceous scales that fit together almost like a suit of mail. The scales of some species also bear needlelike projections. When *Synura* and *Dinobryon,* both colonial genera, bloom in ponds, lakes, and bogs, they impart an odor that is similar to ripe cucumbers or dead fish.

Chromulinales. A number of interesting organisms belong to this order. *Chrysamoeba* has a single chloroplast, narrow delicate pseudopods, and a single small flagellum. The cells of *Chrysarchnion* have interconnected pseudopods, forming a networklike colony.

Craspedomonadales. These forms have a collar or funnel surrounding the single flagellum which emerges at the anterior end of their cells. The current created by the flagellum drives small particles of food toward the anterior cell surface at the base of the funnel where small pseudopods engulf it. Genera that lack chloroplasts are also claimed by the protozoologists, who place them in their order Choanoflagellida. Some genera are colonial. The choanoflagellates are morphologically similar to the choanocytes of sponges, suggesting that choanoflagellates are the ancestral stock from which the sponges evolved.

Dictyochales. These organisms have a siliceous skeleton which consists of two interconnencting hexagonal rings. Although only a few species (e.g., *Dictyocha fibula*) are now abundant in modern seas, the group was quite numerous in the past and it has left a rich fossil record. The protozoologists classify them in an order Silicoflagellida.

The remainder of the chrysophytes (orders Chrysosphaerales, Thallochrysidales, and Phaeothaminales) are coccoid or colonial filamentous forms with motile zoospores.

Haptophyta. (Prymnesiophyceae) As already mentioned, the haptophytes are characterized by a coiled filamentous structure known as a haptonema. Superficial scales produced in the Golgi apparatus cover the surfaces of these organisms. Some have scales made of cellulose; others, the Coccolithophorids, have scales with calcite. The coccolithophorids have a rich fossil record which stretches back to the lower Jurassic period. Modern marine forms include *Hymenomonas carterae, Coccolithus pelagicus,* and *Gephyrocapsa huxleyi.*

Chrysochromulina is another well-known genus which has species with very elaborate scales. *Prymnesium parvum* is a freshwater species that blooms in ponds used for fish culture, producing a fish-killing extracellular hemolytic toxin.

BACILLARIOPHYTA (BACILLARIOPHYCOPHYTA; BACILLARIOPHYCEAE): DIATOMS. Diatoms are abundant in the plankton and in euphotic zones of the benthos of marine and freshwater habitats. They are also found in moist soils, sand, rocks, on plant surfaces, and as endosymbionts. Diatoms have a characteristic siliceous cell wall, known as a *frustrule,* which is built up of two overlapping boxlike halves (valves) between which is a continuous beltlike loop of silica, the *girdle.* The morphology of the frustrule is used to divide diatoms into two orders (centrales and pennales) and for further classification. The scanning electron microscope has been of great aid in the recent study of diatom frustrules.

Diatoms usually reproduce by vegetative reproduction. Each daughter cell receives one parental valve and produces a new, slightly smaller, opposite valve. Although one daughter cell is exactly the same size as the parent, the other daughter is slightly smaller. Thus, in a given diatom population, some cells are undergoing progressive diminution in cell size. Restitution to full cell size may take place by the extrusion of the protoplast (cell contents bounded by a cell membrane) from a small frustrule and the regeneration of a new frustrule, or by sexual reproduction. In centric diatoms (centrales), the gametes are sexually differentiated (oogamous) but in pennate diatoms, they are alike (isogamous). The zygote is an auxospore that increases in volume to produce large-sized vegetative cells. Most benthic and neritic centric diatoms also produce resting spores that sink to the bottom, where they persist during unfavorable conditions. Vegetative reproduction can lead to the formation of filamentous chains or flattened ribbons. Some diatoms

Algae

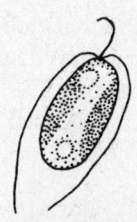

Prymnesium
∼ X 600

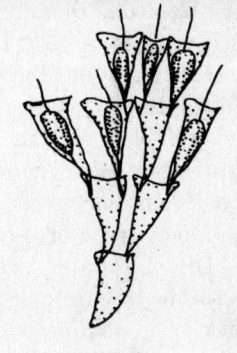

Dinobryon
∼ X 1,000

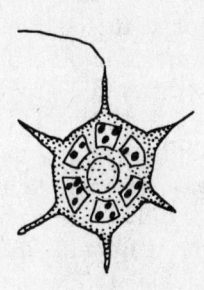

Dictyocha
∼ X 600

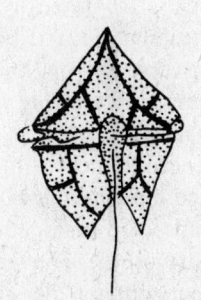

Peridinium
∼ X 400

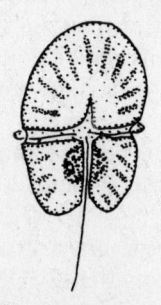

Gymnodinium
∼ X 400

Porphyridium
∼ X 3,000

(*Schizoema, Berkeleya*) form tubular colonies that reach macroscopic sizes.

Pyrrhophyta (Pyrrhophycophyta, Protozoan Order Dinoflagellida): Dinoflagellates. Dinoflagellates are a diverse group distributed widely in marine brackish and fresh bodies of water. They have distinctive nuclei (*mesocaryotic*) with permanently condensed beaded chromosomes that lack histones, a feature not found in any other group of algae or protozoa.

Dinoflagellates have chlorophylls a and c, diadinoxanthin, dinoxanthin, β-carotene, and peridinin. Some forms produce trichocysts. The phylum consists of two major classes: the Desmophyceae and the Dinophyceae. The Desmophyceae have two flagella originating from their anterior ends and are divided into right and left valves. The Dinophyceae have a pair of heterodynamic flagella which arise laterally. One flagellum, the transverse flagellum, is pleuronematic and bandlike, and is wrapped around the body in a transverse furrow known as the *cingulum* or *girdle*. The other flagellum, the *trailing flagellum,* is acronematic and lies in a groove, the *sulcus*.

The outer covering of dinoflagellates is a multilayered structure called an *amphiesma,* which includes the plasma membrane. Some groups of dinoflagellates are *thecate* (armored), others are *naked* (unarmored). The armor of thecate dinoflagellates is built up of *thecal plates*. The simplest theca consists of two plates (*Prorocentrum*) but in many groups there are 20 or more plates. Some thecate dinoflagellate genera have cysts, produced within the parental wall, which are also built of thecal plates. The remains of armored dinoflagellates are also found in the fossil record.

Dinoflagellates commonly reproduce asexually by longitudinal cell division while they are still swimming. Cleavage lines in armored forms follow plate patterns. The daughter cells produce new plates to replace missing ones. Sexual reproduction, both isogamous and anisogamous, has been described in a number of genera. In *Ceratium* spp. the male gamete is much smaller but otherwise similar to the female vegetative cell. In some genera the gametes are morphologically similar to the genus *Gymnodinium*.

Many dinoflagellates (e.g., *Noctiluca, Gonyaulax, Pyrocystis, Pyrodinium*) are bioluminescent. When seawater containing these organisms is disturbed by wind and wave action or by the wake of a passing ship, the greenish glow is visible for considerable distances.

Blooms of dinoflagellates, or *red tides,* contain millions of cells per liter. Certain of these blooming genera (e.g., *Prorocentrum, Gymnodinium, Gonyaulax, Ceratium*) have species that produce powerful toxins. Some toxic species (e.g., *Gymnodinium breve*) produce blooms that kill fish but few invertebrates. Other species (e.g., *Gonyaulax catenella*) produce a powerful toxin that is transferred and concentrated in clams, mussels, oysters, and scallops when they feed upon the dinoflagellates. The toxin, known as saxitoxin, is concentrated in the siphon or digestive gland of the shellfish. It is a neurotoxin which prevents normal transmission across neuromuscular synapses.

Although the majority of dinoflagellates are free-living forms, some are symbionts or parasites. The best known of these is *Zooxanthella microadriaticum* (formerly *Symbiodinium microadriaticum*), an endosymbiont in a wide variety of marine invertebrates including corals, jellyfish, sea anemones, foraminifera, and the giant clams (*Tridacnia*). These algae appear to provide their hosts with carbohydrates and greatly enhance calcification in corals and foraminifera.

Cryptophyta (Cryptophycophyta, Protozoan Order Cryptomonadida): Cryptomonads. This is a small phylum of red, blue, olive-yellow, brown, or green marine and freshwater algae with palmelloid and coccoid forms, and having two unequal flagella. They have chlorophylls a and c, diadinoxanthin, dinoxanthin, α- and β-carotene, zeaxanthin, phycocyanin-c, phycoerythrin-ε and other phycobilin red pigments similar to those of the Rhodophyta and the cyanobacteria. Their reserve food is starch. Unlike other algal groups, both flagella are pleuronematic. The flagella arise ventrally from within a depression or furrow, which opens close to the anterior end of the cell. Cryptomonads have coiled, dischargeable organelles, *ejectosomes,* lining their gullet. Ejectosomes are analogous to the trichocysts of dinoflagellates and ciliates. The best-known genera are *Cryptomonas* and *Chroomonas. Chilomonas paramecium* is a well-known nonpigmented species.

Rhodophyta (Rhodophycophyta): Red Algae. Because there are only a few (eight) genera of unicellular rhodophytes (e.g., *Porphyridium*), this large group of seaweeds is of only marginal interest to microbiologists. Rhodophytes are distinguished from other algal groups by the absence of any flagellated stages, the characteristic ultrastructure of their chloroplasts, the storage of floridean starch, and the presence of large amounts of the water-soluble phycobilins (phycoerythrin and phycocyanin) which mask the presence of chlorophylls a and d. They reproduce both asexually and sexually.

FUNGI

Fungi, commonly known as the molds and yeasts, are a distinct kingdom of chemoorganotrophic eucaryotes that lack chlorophyll, have a rigid cell wall, and typically, have a branched, tubular vegetative structure, the *thallus,* that is capable of almost indefinite growth. The thallus is composed of individual filaments called *hyphae* (singular, *hypha*). Groups or masses of intertwining branching hyphae are known as *mycelia* (singular, *mycelium*). The thallus is differentiated into vegetative mycelia and reproductive mycelia. Hyphae that penetrate the substrate and anchor the thallus are known as *rhizoids.* Parasitic and lichen symbiotic fungi produce special hyphae, *haustoria,* which penetrate the cells of the living host to obtain nutrients.

The hyphae of various groups of fungi differ internally. Some (e.g., those of the Phycomycetes) are *nonseptate* (coenocytic, acellular), that is, they form no internal cross walls (*septa*) after nuclear divisions except at the base of reproductive structures. In this type of mycelium, there is extensive flow of cytoplasm and nuclei, particularly within the growing hyphal tips. Another fungal group (Ascomycetes) has *septate hyphae* with uninucleate cells. The septum usually has a central pore that permits the flow of nuclei and cytoplasm from one cell to another. Some septate mycelia have multinucleate cells.

Although most fungi are filamentous, many are unicellular or colonial, making the boundaries of the kingdom somewhat arbitrary; three classes and part of another class of lower fungi (Acrasiomycetes, Protosteliomycetes, Myxomycetes) are also considered by protozoologists to be protozoan classes (Acrasea, Eumycetozoea, Plasmodiophorea), and one group is a protozoan phylum (Labyrinthomopha).

The classification of fungi is based largely on the characteristics of their fruiting bodies and of their sexual and asexual spores. Regardless of the system of classification used for the overall separation of living organisms into kingdoms (see Chapter 1), the fungi fall into at least two natural groups: the Eumycota (true fungi) and the Myxomycota (the true slime molds, which overlap the boundaries of the protozoa.) The Eumycota include the Ascomycota, Basidiomycota, and Zygomycota. In the Whittaker five-kingdom scheme, the Oomycota and the Chytridiomycota are included in the true fungi. However, in Leedale's scheme, the chytrids are a separate kingdom and the Oomycota is a phylum in the heterokonts. Fungi whose sexual stages are unknown are provisionally grouped together as the Fungi Imperfecti, or form-class Deuteromycetes.

Spores. Spores ensure the survival of fungal species between optimum growth conditions and are the means by which they are broadcast to favorable new environments. There are two kinds of spores: asexual and sexual.

ASEXUAL SPORES. Asexual spores are not dormant structures, nor are they heat resistant; they are, however, resistant to drying and radiation, and germinate when moisture becomes available. Quite a variety of asexual spores are produced by different fungal groups.

Chlamydospores and Arthrospores (Oidium). These are the simplest types of spores. Chlamydospores are produced when some of the individual cells of the mycelium become rounded and produce thick walls. They then detach from the rest of the mycelium. Arthrospore formation is similar except that the spores have less food reserve and they are rectangular fragments of the parent hyphae.

Sporangiospores. This type of spore is produced within a saclike structure, the *sporangium* (plural, *sporangia*), which is borne at the tip of a specialized hypha, the *sporangiophore*. When the sporangium ruptures, it releases one of two types of sporangiospores, *zoospores* (flagellated and motile) or *aplanospores* (nonmotile).

Conidiospores (Conidium, Plural, Conidia). These spores are also formed at the tip of a specialized hypha, the *conidiophore*. However, instead of being enclosed, conidiospores are formed free, sometimes in long chains, like a string of beads. There is a wide range in the morphology of conidia, a feature that is useful in taxonomy. Some genera produce two types of conidia on the same thallus: microconidia and macroconidia. Microconidia are small and usually unicellular; macroconidia are larger, multicellular.

Blastospores. Blastospores are thin-walled cells budded from pseudohyphae (temporary colonies of yeasts).

Spores of Rusts or Smuts. Rusts or smuts (class Basidiomycetes, subclass Teliomycetidae) are plant pathogens with complex life cycles involving the formation of several different types of asexual spores. *Aeciospores* are haploid but binucleate (dikaryotic) spores produced in chains from the base of a cup-shaped organ formed within the epidermis of the host plant. *Urediniospores* are budded from a mycelium growing subepidermally within the host. As the dikaryotic spores form, they press against the inside of the epidermis, eventually rupturing it, and release themselves. *Teliospores* are thick-walled, overwintering, binucleate spores produced by rusts. Eventually nuclear fusion (karyogamy) takes place, turning each spore into a zygote.

SEXUAL SPORES. Sexual spores are formed after the fusion of unicellular gametes.

Oospores. Oospores are produced in the Oomycota by the fusion of flagellated sperm (formed in a specialized organ, the *antheridium*) with an *oosphere* (egg, produced within a specialized organ, the *oogonium*).

Zygospores. Zygospores are produced in the zygomycota by the fusion of the nuclei in two *gametangia* which have joined together during conjugation. A gametangium (sex cell) is formed at the tip of each hypha by the formation of a cross wall when two mycelia of opposite sex come in contact with one another.

Ascospores. Ascospore formation is characteristic of the Ascomycota. A number of different mechanisms (e.g., gametangial copulation; gametangial contact; spermatization; somatogamy) have evolved within this large group to achieve the union of two compatible nuclei. Regardless of the method by which the nuclei are brought together, nuclear fusion takes place in the ascus mother cell. Almost immediately after fusion, the zygotic nucleus undergoes two meiotic divisions and a mitotic division to give rise generally to eight ascospores. In most of the classes, the *ascus* (plural, *asci*), a saclike structure that contains the ascospores, is enclosed in a specialized fruiting body (*ascocarp*). Ascocarps vary greatly in the different groups of Ascomycetes.

Basidiospores. Basidiospores develop at the end of a club-shaped structure, the *basidium* (plural, *basidia*) in the Basidiomycota. The basidium originates as the terminal cell of a binucleate hypha. After the nuclei fuse (karyogamy), the zygote undergoes two meiotic divisions which give rise to the four basidiospores that are typcially borne on each basidium. The more complex basidiomycots produce basidia in fruiting bodies (basidiocarps) which are familiar to us as mushroom caps, puffballs, and so forth.

Chytridiomycota: Chytrids. The chytridiomycota are distinguished from other fungi by the production of posterior uniflagellated zoospores and planogametes. The chytrids are primarily aquatic, but they also inhabit the soil. Many are parasites. *Synchytrium endibioticum,* the organism that causes the disease known as potato wart, is perhaps the most destructive and well-known parasite in the group. *Olpidium brassicae* infects cabbage roots. *Physoderma Zae maydis* causes brown spots on corn. *Allomyces* and *Blastocladiella* are aquatic species that have been studied extensively in the laboratory. The female gametangia and gametes of *Allomyces* produce a sexual hormone, *sirenin,* which is an attractant for male gametes. The life cycles of the chytrids involve alternation of haploid (haplophase) and diploid (diplophase) generations. Flagellated zoospores or zygotes produced by the copulation of flagellated planogametes are usually involved in the infection of the host.

Acrasiomycota [Protozoan Classes Acrasea and Eumycetozoea (in part)]: Cellular Slime Molds. These slime molds form a pseudoplasmodium (grey slug) in which the component myxamoebae do not fuse but retain their individuality. The life cycle involves the germination of spores, each of which releases a single uninucleate haploid amoeba. The individual amoebae feed on bacteria, grow, and divide by binary fission. When the population reaches a critical size, the amoebae cease feeding and aggregate under the influence of a hormone (acrasin) to form a *pseudoplasmodium.* The pseudoplasmodium may migrate for some distance before it produces a stalked fruiting body (*sorocarp*) which in turn produces spores, thus completing the asexual phase of the life cycle. The sexual phase of the life cycle involves fusion of two myxamoebae to form a zygote (*macrocyst*). Meiosis takes place and myxamoebae are released upon germination of the macrocyst. The best-known genera of cellular slime molds are *Dictyostelium, Polyspondylium,* and *Acrasis.*

Oomycota. The Oomycota differ from other fungi in that their zoospores are biflagellate with a pleuronematic flagellum directed forward and an acronematic flagellum directed backward. Their sexual reproduction is oogamous by gametangial contact. Meiosis takes place in the oogonium (egg-producing organ) and antheridium (sperm-producing organ) just before the gametes are formed. One nucleus of the antheridial cell passes through the fertilization tube into each oogonium and fuses with an egg nucleus. Members of the group are either free living or parasitic on algae, water molds, small animals, and terrestrial plants. *Phytophthora infestans,* the cause of the Irish late blight of potatoes, is exceptionally well known. *Saprolegnia* species cause diseases of fish and fish eggs. *Aphanomyces* species cause serious disease of sugar beets, peas, and other crops. Downy mildews caused by *Plasmopara viticola* have caused destruction of vineyards.

Labyrinthulomycota (Protozoan Phylum Labyrinthomorpha). This is a distinct group of fungi with characteristics unlike those of any other group. They do not form hyphae, pseudoplasmodia, or orthodox pseudopodia. Individual trophic cells are spindle-shaped or spherical and move through channels of slime secreted by the cells. The individual slime filaments are thin and form branching and anastomosing networks. They produce biflagellated heterokont zoospores in sori or pseudosori from single cells. Labyrinthulids parasitize sea grasses, seaweeds, and some unicellular algae (diatoms and chlorophytes). There is some evidence to suggest that *Labyrinthula marina* was responsible for the "wasting disease" of eel grass.

Myxomycota (Protozoan Subclass Myxogastria). These are the true slime molds (acellular slime molds). Their distinguishing feature is a free-living (they eat bacteria), acellular, multinucleate plasmodium which behaves as a functional unit at all times and eventually gives rise to a fruiting body (sporophore) in which meiosis takes place and spores are produced. The spores give rise to myxamoebae, or flagellates (most have two flagella, a few only one), which fuse in pairs to form zygotes, completing the life cycle. Under certain environmental conditions, the plasmodia of some species are converted into a hardened mass (*sclerotium*) that can remain dormant for some time (e.g., over winter). When conditions become favorable again, the plasmodium grows out of the sclerotium. Acellular slime molds usually live in the shade on decaying logs, dead leaves, tree bark, plant debris, animal dung, and some grasses. They are of little economic importance. Occasionally they creep over ornamental plants or destroy lawns. The best-known species is *Physarum polysephalum,* an organism that has been used in many molecular and cellular biological studies.

Zygomycota. The Zygomycota produce a thick-walled resting spore (*zygospore*), the result of the complete fusion of two gametangia (hyphae that contain gametic nuclei). Most zygomycetes

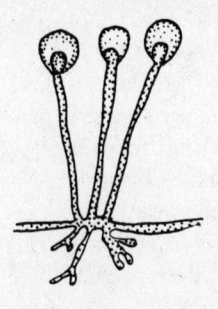

Portion of a hypha of
Rhizopus with sporangia
∼ X 10

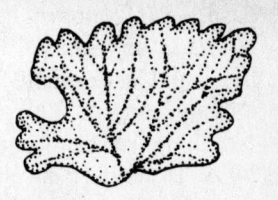

Phaneroplasmodium
of *Physarum*
∼ X 20

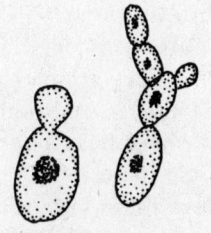

*Saccharomyces
cerevisiae*
with buds
∼ X 100

Portion of a
hypha of *Penicillium*
with a conidiophore
∼ X 20

Perithecium
of *Neurospora*
with asci
∼ X 10

have well-developed coenoytic mycelia. Motile cells are not produced in any phase of the life cycle. Sporangiospore formation is the usual asexual method of reproduction. Some species produce chlamydospores. The majority of organisms in this group are saprobic, living on decaying plant or animal matter, dung, food, and so forth. Others are carnivorous, parasitic (even parasites of other fungi), and mycorrhizal. Many of the saprobic species are easily cultured and are used commercially to synthesize important industrial products (e.g., fumaric acid, cortisone, alcohol, lactic acid, citric acid, oxalic acid). A number of species ruin crops. For example, *Rhizopus stolonifera* ruins strawberries in transit, and sweet potatoes in storage. It is also a common bread mold. *Choanephora cucrubitarum* attacks squash blossoms and fruit. The majority of species in one interesting order, the Zoopagales, traps and eats amoebae and nematodes. The mycelia of such predatory fungi are coated with sticky substances that capture and hold the prey. Fine branched haustoria invade the animal and absorb nutrients. The organisms in one large class, the Trichomycetes, are symbionts in the hind gut of insects, millipedes, and crustaceans. They attach to the chitinous gut lining but do not penetrate host tissues. Asexual reproduction in this particular group produces amoeboid cells, arthrospores, and highly modified sporangiospores (trichospores).

Ascomycota. The Ascomycota produce ascospores following karyogamy (fusion of gametic muclei). As mentioned earlier in this chapter, ascospore formation is preceded by meiotic nuclear division and one or more mitotic divisions. The mycelia of Ascomycetes are septate and their walls are generally rich in chitin.

Some Ascomycetes are commercially important. The fermentative abilities of yeasts are the basis for the baking and brewing industries. Morels and truffles are gourmet delicacies. Deadly, but medicinally important, alkaloids are produced by the ergot fungus, *Claviceps purpurea,* when its mycelium invades and destroys the ovaries of rye and other grasses. Various species of Ascomycetes destroy crops and timber, causing disease such as apple scab, brown rot, powdery mildews, corn ear rot, chestnut blight, and Dutch elm disease.

An ascomycete life cycle usually has two phases: the sexual (ascigerous or perfect) stage and the asexual (conidial or imperfect) stage. Some species do not have conidial stages or they are still undiscovered.

Ascomycetes are divided into five classes on the basis of mycelial and ascocarp morphology.

CLASS HEMIASCOMYCETES (PROTOASCOMYCETIDAE). These are fungi that form asci directly from zygotes without forming ascogenous hyphae or ascocarps. As a group they are quite abundant on the surfaces of fruit, nectar of flowers, and substrates with fermentable sugars. *Saccharomyces cerevisiae* is the familiar yeast used in the brewing and baking industries. *Taphrina deformans* is the agent that causes peach and almond leaf curl disease.

CLASS PLECTOMYCETES. The Plectomycetes have thin-walled asci that produce unicellular ascospores.

CLASS ELAPHOMYCETALES. The Elaphomycetales are the truffle fungi that form their ascocarps under the ground. The strong odor emitted by the ascocarps attracts deer and pigs. *Arthroderma* spp. and *Nannizzia* spp. cause the skin diseases popularly known as ringworm of the scalp, body, beard, feet, or nails. *Ajellomyces dermatitidis* causes a disease known as blastomycosis. *Emmonsiella capsulata* causes histoplasmosis, a widespread disease that is sometimes fatal to humans. *Aspergillus flavus* and other *Aspergillus* spp. grow in a variety of foods, causing decay and the production of toxins (aflatoxins). *Aspergillus flavus, A. fumigatus,* and *A. niger* also cause diseases collectively known as *aspergilloses*. The symptoms of the lung aspergillosis resemble those of tuberculosis. Species of the blue or green mold *Penicillium* are widespread, causing decay of fruit, leather, and fabrics. *Penicillium* species are important in the manufacture of cheese (e.g., Roquefort, Camembert, Gorgonzola) and antibiotics (e.g., penicillin, griseofulvin).

CLASS PYRENOMYCETES. These forms produce club-shaped or cylindrical asci in an enclosed ascocarp (perithecium) with a pore (ostiole) at the top. One family (Erysiphaceae) produces the plant diseases known as powdery mildews. Several fungi in the order Xylariales cause serious diseases of plants. *Rosellina nectrix* is responsible for root rot of grapevines. Other species cause various cankers on trees. Not all pyrenomycetes are parasitic. Many live on decaying plant litter, logs and stumps, dung, etc. The best known of this latter group of fungi is the red bakery mold, *Neurospora crassa*, which has been employed widely as a genetic tool.

CLASS DISCOMYCETES. These Ascomycetes form open ascocarps. The group includes cup fungi, earth tongues, morels (sponge mushrooms), and truffles. Their spores are puffed out in clouds. All the true morels (*Morchella* spp.) are edible and considered by epicures to be delectable.

Fungi Imperfecti. The imperfect fungi (Deuteromycotina, form-class Deuteromycetes) are fungi with septate hyphae that reproduce only by means of conidia. The group is known as imperfect because the organisms included either have lost the sexual phase in their life cycle or have a sexual stage that has not yet been found. Most of the organisms placed in the Deuteromycetes are very similar to the conidial stages of well-known Ascomycetes. The group is large (more than 15,000 species). Most are saprobes or weak plant parasites in terrestrial habitats but a small proportion cause serious diseases of plants, animals, and humans. *Rhodotorula, Cryptococcus, Candida,* and *Torulopsis* are better known genera of yeasts, and are routinely isolated from fresh and marine waters, soil, plant litter, and animals. *Candida albicans,* part of the normal flora of mucous membranes of many mammals, is also a pathogen causing cutaneous candidiasis, oral candidiasis (thrush), bronchiocandidiasis, and vulvovaginal candidiasis. *Trichosporon cutaneum* causes white piedra, an infection of hair on the scalp, beard, mustache, and pubic region.

A number of Deuteromycetes are important plant pathogens. *Fusarium oxysporium* produces wilt on important crop plants such as bananas, tomatoes, sweet potatoes, and pears. *Cercospora* species cause leaf spot diseases on celery, beet, lettuce, tomato, potato, and tobacco.

Basidiomycota. As in the case of the large seaweeds, most of the groups of Basidiomycota are too large to be of more than casual interest to microbiologists. Mushrooms, toadstools, puffballs, stinkhorns, shelf fungi, jelly fungi, and bracket fungi are well-known Basidiomycetes. The rusts and smuts are small enough to fall into the general microbial realm. Basidiomycetes differ from other fungi in the formation of a *basidium,* a swollen, club-shaped, sexually reproductive cell which is produced by a binucleate cell at the tip of a hypha. The two nuclei fuse, forming a zygote that undergoes meiosis to form four haploid nuclei. Up to this point the process is similar to the sexual aspect of Ascomycetes reproduction. The next step in the Ascomycetes life cycle would be the formation of ascospores. In the Basidiomycetes, however, the haploid nuclei are extruded to the tips of four projections (*sterigma*) at the end of the basidium where a thick spore wall is formed around them. In the mushrooms the basidia and basidiospores are formed along the underside of the familiar fleshy umbrella-shaped fruiting bodies.

CLASS TELIOMYCETES (SUBCLASS TELIOMYCETIDAE OF SOME AUTHORITIES). This group includes the rusts, smuts, and the basidiomycetous yeasts. Characteristically Teliomycetes produce a thick-walled binucleate resting spore (*teliospore*). Karyogamy (nuclear fission) takes place within the teliospores. When teliospores germinate, they give rise to a germ tube (promycelium) in which meiosis takes place. Meiosis is followed directly by basidiospore formation.

The Teliomycetes are divided into two orders, the Uridinales (rusts) and the Ustilaginales (smuts and basidiomycetous yeasts).

Uredinales. The rusts are plant parasites that cause economically important diseases including black stem rust of cereals, coffee rust, asparagus rust, carnation rust, cedar-apple rust, and snapdragon rust.

The life cycles of rusts are complex. Besides teliospores and basidiospores, other types of spores are produced by most rusts as well (*Aecia* bear *aeciospores, uredinia* bear *urediniospores*). Perhaps the best known of the rusts is *Puccina graminis,* which grows on wheat, barley, rye, and oats. The two-celled diploid teliospores are produced on the leaves and stems of the grain during the overwintering stage. The following spring a promycelium develops from each cell wall which in turn produces basidiospores. The basidiospores are carried by the wind to barberry bushes where they germinate. A mycelium develops on the barberry which produces spermogonium with spermatiophores. Flies or other insects attracted by the fragrances of the spermogonal mass transfer spermatia from one spermogonium to the receptive hyphae of the next. If the spermatia are of the appropriate mating type, the spermatial contents pass into the receptive organ at the point of contact. The spermatial nuclei pass to aecial primordial cells, rendering them binucleate. The aecial primordial cells develop into aecia which produce binucleate (dikaryotic) aeciospores. The aeciospores break through the epidermis of the barberry, and are disseminated by the wind. Germination, which occurs on a grass host, leads to the development of a binucleate mycelium which forms masses of cells (uredinia). From these, in turn, are formed rusty red urediniospores. The urediniospores germinate and produce new mycelia and new crops of urediniospores. At the end of the summer the uredinia produce telia and teliospores, thus completing the cycle. One of the methods of eradicating the disease is the removal of barberry bushes near wheat fields to break the life cycle.

Ustilaginales. The smuts are also plant pathogens causing such economically important disease as corn smut, loose oat smut, and onion smuts. The life cycle of *Ustilago maydis,* the causative agent of corn smut, is most often used for representation of the group. In common with the rusts, the teliospore of smuts can be an overwintering stage. When the smut teliospore germinates, it also produces a promycelium in which meiosis takes place. Septa are formed which then divide the protoplasm and lead to the production of basidiospores. Germination of the basidiospores leads to the formation of a uninucleate mycelium that infects the host. If hyphae of compatible strains meet in the host, they fuse and the resulting mycelium becomes dikaryotic (binucleate). The binucleate mycelium produces branches that reach the surface of the host. There they give rise to several crops of binucleate conidia and conidiospores. Disseminated by the wind, conidiospores initiate new infection later in the growing season of the host. They produce smut balls (sori) containing teliospores.

PROTOZOA

Protozoa is the collective taxonomic group used by zoologists to encompass all the unicellular and simple colonial animals. Contemporary protozoologists view the group they study as a kingdom or subkingdom of animals with seven phyla. As mentioned in the discussions of algae and fungi, it is not easy to draw clearly defined boundaries between these major groups and the protozoa. Because of this, many groups of organisms are included taxonomically in more than one kingdom. For example, almost every order placed in the protozoan class Phytomastigophorea is considered to be a phylum by phycologists (scientists who study algae). The organisms in each of these groups (phyla or orders) range from species with typical algal photosynthetic and storage organelles to forms that are colorless and holozoic (ingest organic matter) or saprozoic (absorb dissolved nutrients). The transition between the extremes is gradual, and many species have both photosynthetic capacity and holozoic or saprozoic abilities. Four protozoan classes (Acarpomyxea, Acrasea, Eumycetozoea, and Plasmodiophorea) and one phylum (Labyrinthomorpha) are considered various forms of slime molds by mycologists (scientists who study molds). Confusing as it may be to encounter the same organism in more than one taxonomic group, dual classification is to be expected from time to time. It must be remembered that taxonomic schemes are just means devised by scientists to sort out and organize information about the living world and that allowance has to be made for differing points of view.

While most protozoa range in size from 5 to 250 μm, some species are as small as 1 μm and others as large as 12 cm. Approximately one-third of the 35,000 species of living protozoa are parasitic. Some species and large groups of protozoa are well-known endosymbionts (e.g., rumen ciliates, termite flagellates); conversely, some protozoa serve as hosts for endosymbionts (e.g., *Paramecium bursaria* and larger foraminifera). The Foraminiferida, Spumellarida, Tintinnina, and Nassellaria have a long and rich fossil record. More than half of the 65,000 named species of protozoa are fossils. As is true with all other groups of microorganisms, electron microscopy has had a major impact on recent taxonomic changes in the protozoa. The criteria that separate the seven phyla of protozoa (Sarcomastigophora, Labyrinthomorpha, Apicomplexa, Microsporea, Ascetospora, Myxozoa, and Ciliophora) are based on the subcellular specializations in organelles that function in locomotion, feeding, and reproduction.

Phylum Sarcomastigophora. This is the largest of the protozoan phyla. As its name implies, the organisms in this group move by flagella and pseudopodia; in rare instances they have the ability to move by both means, although not necessarily in the same phase of the life cycle. Typical members of the phylum have a single type of nucleus (as opposed to differentiation into micro- and macronuclei) and do not form spores. Sexual reproduction is known to take place in many but not all of the subgroups. The phylum is subdivided on the basis of locomotory means into three subphyla: Mastigophora, Opalinata, and Sarcodina.

SUBPHYLUM MASTIGOPHORA. Members of this subphylum have one or more flagella and typically reproduce asexually by binary fission. Flagellates with chloroplasts, and those clearly related to chloroplast-bearing forms, are assigned to the class Phytomastigophorea. The remaining flagellates are placed in the class Zoomastigophorea. Most of the Phytomastigophorea are free living. Most, but not all, of the zooflagellates are parasites or endosymbionts.

Class Phytomastigophorea. Each of the orders of the Phytomastigophorea (commonly called phytoflagellates) corresponds to all or part of an algal kingdom, division, phylum, class, or order. Not all of the algae, however, are accepted by protozoologists for inclusion in their group. The correspondence of the protozoan orders to the algae are given in Table 13.1.

Class Zoomastigophorea. The zoomastigophorea is a polyphyletic group. The various orders are presumed to have evolved from phytoflagellate ancestors.

<u>Order Choanoflagellida.</u> Choanoflagellates are an interesting group. Each organism has a single apical flagellum surrounded by a ring of tentacles or a collar. The flagellum creates a current which drives small food organisms to the surface of the cell within the collar where they are ingested. Many of these "collared flagellates" are stalked and attached. A number have protective coverings

Table 13.1. Correspondence of Protozoan Orders to Algal Groups

PROTOZOAN ORDER	ALGAL GROUP
Cryptomonadida	Cryptophyta (kingdom or division)
Dinoflagellida	Pyrrophyta (kingdom or division) or Dinophyta
Euglenida	Euglenophyta (kingdom or division)
Chrysomonadida	Chrysophyta (phylum or division of some authorities) or Chrysophyceae (class of other authorities)
Chloromonadida	Chloromonadophyceae (class)
Heterochlorida	Xanthophyta (only part of this division or phylum)
Prymnesiida	Haptophyta (kingdom or phylum of some authorities) or Prymnesiophyceae (class of other authorities)
Volvocida	Volvocales (order)
Prasiniomonadida	Prasiniophycales (order) or Prasiniophyceae (family)
Silicoflagellida	Dictyochales (order)

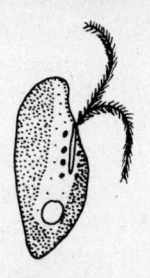

Protozoa

Cryptomonas
~ X 2,000

Pleurosiga
~ X 1,000

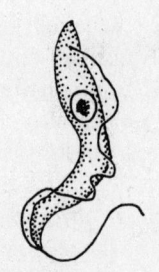

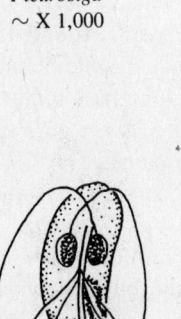

Trypanosoma
~ X 3,000

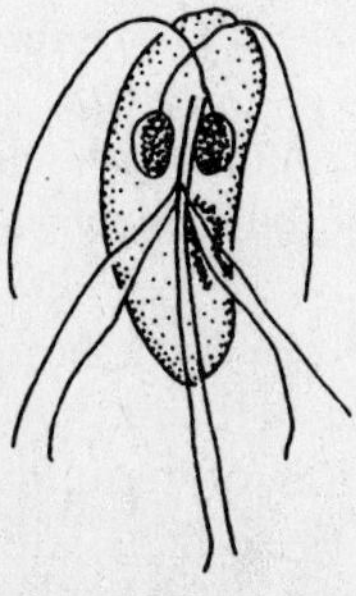

Giardia
~ X 2,000

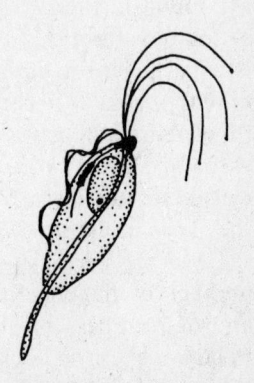

Trichomonas
~ X 1,500

called *loricae* composed of siliceous costae. The zoomastigophorean order corresponds to the nonpigmented forms in the algal order Craspedomonadales. (Phylum Chrysophyta).

Order Kinetoplastida. These forms have one or two flagella arising from a depression, and a characteristic organelle, the *kinetoplast*. The kinetoplast is a conspicuous DNA-containing (feulgen-positive) body located at the end nearest the flagellar kinetosomes of the rather large single sausage-shaped, hoop-shaped, or branched mitochondrian which characteristically extends the entire length of the body. The Golgi apparatus is also typically located in the region of the flagellar depression. The best-known members of this order are members of the genera *Trypanosoma* which infect all classes of vertebrates. The trypanosomes of fish are transmitted by leeches, those of terrestrial vertebrates by biting insects (sand flies, tsetse flies, etc.). Chagas' disease, an often fatal disease in South and Central American children, is caused by *Trypanosoma cruzi*.

Order Diplomonadida. Diplomonads characteristically have multiple body organelles. The best-known genera, *Hexamita* and *Giardia*, are bilaterally symmetrical and have two nuclei and two sets of four flagella. *Giardia lamblia* is a human intestinal parasite causing diarrhea, chronic stomach ache, nausea, and so on. *Hexamita meleagridis* causes enteritis in turkeys.

Order Trichomonadida. The trichomonads have three to five anterior flagella and one recurrent trailing flagellum arising out of the anterior of the cell. A stiff central rod (axostyle) cups around the nucleus and emerges from the posterior of the cell. Trichomonads are found in the digestive tracts, cloacae, and urinogenital systems of many vertebrates and some invertebrates. *Trichomonas gallinae* is a pathogen in the anterior digestive tract of pigeons. *Tritrichomonas foetus* is a parasite of the genital tract of cattle and sometimes causes abortion. Three species of *Trichomonas, T. tenax, T. vaginalis,* and *Pentatrichomonas hominis,* parasitize humans. None causes serious disease. *Trichomonas vaginalis* is the best known of the three; it causes "vaginal itch."

Order Hypermastigida. This group is characterized by a large number of flagella, multiple axostyles, and parabasal bodies. All known members of the order are intestinal parasites of termites, wood roaches, or cockroaches. The best known of these flagellates belong to the *Trichonympha, Spirotrichonympha,* and *Barbulanympha*.

Orders Protermonadida, Retortomonadida, and Oxymonadida.
These orders contain less well-known parasitic flagellates.

SUBPHYLUM OPALINATA. The forms in this subphylum superficially resemble ciliates because they are covered with simple oblique rows of flagella. They differ, however, in cortical organization and form of division, and their nuclei are not differentiated into micro- and macronuclei. Except for a few species that have been found in fish and snakes, opalinids are parasites in the large intestines of amphibians.

SUBPHYLUM SARCODINA. These are the amoebas, characterized by the presence of pseudopodia. The gametes of some species and temporary or developmental stages of others may be flagellated. Sexuality is not known in every order. Two superclasses are recognized; the Rhizopodea, with relatively simple pseudopodia, and Actinopodea, with long, thin pseudopodia stiffened by microtubules.

SUPERCLASS RHIZOPODEA. This superclass encompasses forms with simple pseudopodia. These animals have lobose pseudopodia, are usually uninucleate, and do not form sorocarps, sporangia, or other similar fruiting bodies.

Order Amoebida. The forms in this order are typically uninucleate. *Amoeba proteus* is the best-known species in the group. The most important parasitic amoeba of humans is *Entamoeba histolytica*, the cause of amoebic dysentery.

Order Schizopyrenida. An amoeba in this order is basically a cylinder with a single eruptive pseudopodium. Most species have temporary flagellate stages. The best-known genera are *Naegleria, Tetramitus,* and *Vahlkampfia. Naegleria* is an interesting small soil amoeba which has, on occasion, been associated with fatal invasions of the human central nervous system.

Order Pelobiontida. Members of this order are also basically cylindrical with a single pseudopodium. They are typically multinucleate and lack mitochondria. Most species examined have endosymbiotic bacteria. The best-known genus is *Pelomyxa*.

Order Arcellinida. These organisms form a test or tectum which may be totally organic or composed of inorganic materials cemented into an organic matrix. The best-known genera are *Arcella* and *Difflugia.*

Order Trichosida. This order was introduced to accommodate a single genus of marine amoebas, *Trichosphaerium.* They have a fibrous sheath which, at one stage of the life cycle, has calcareous spicules.

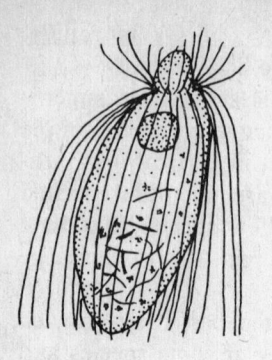

Trichonympha
~ X 300

Opalina
~ X 200

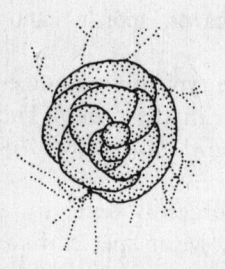

Discorbis
~ X 75

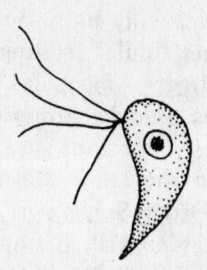

Tetramitus
~ X 3,000

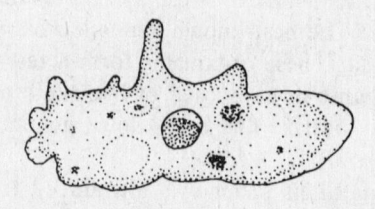

Amoeba
~ X 1,000

Class Acarpomyxea. These amoebas form small branching plasmodia which sometimes become a reticulum of coarse branches. There is no regular reversal of streaming in the plasmodium. These amoebas do not form fruiting bodies or spores. There are two orders in this class: the Leptomyxida, which typically form thin sheets in soils or fresh waters, and the Stereomyxida, which are marine organisms with branched pseudopodia. None of the genera in this class is well known to nonspecialists.

Class Acrasea. This class contain cellular slime molds; it corresponds to the fungal phylum Acrasiomycota (p. 224).

Class Eumycetozoea. The three subclasses of the protozoan class Eumycetozoea (Protosteliia, Dictyosteliia, and Myxogastria) correspond to the two fungal classes Protosteliomycetes and Myxomycetes (page 224).

Class Plasmodiophorea. This class consists of obligate intracellular parasites with minute plasmodia. None of the species form spores or flagellated stages. There are two orders: the Aconchulinida, which contains organisms that lack an external skeleton, and the Gromiida, which has organisms whose bodies are enclosed by a test. The best-known genera in this class are *Vampyrella, Euglypha,* and *Gromia.*

Class Granuloreticulosea. The organisms in this class have delicate, finely granular reticulopodia. There are three orders, Athalamida, Monothalmida, and Foraminiferida. The first two orders are small and have only a few genera, but the third, the Foraminiferida, contains more than 35,000 fossil and living species. The Foraminifera have a single or multichambered test. Various suborders have organic, agglutinated, or calcareous tests. The classical life cycle of the Foraminifera is an alternation of a uninucleate, haploid, sexually reproducing generation with a multinucleate, diploid, asexually reproducing generation (Fig. 13.2). Some species have nuclear dimorphism, with a number of generative nuclei and one (or more) somatic nuclei. Some of the best-known genera are *Allogromia, Discorbis, Quinqueloculina, Elphidium,* and *Ammonia.*

Class Xenophyophorea. This class contains a number of poorly known marine amoebas which form a multinucleate plasmodium enclosed in a transparent branched-tube system. Numerous barite crystals are found in the cytoplasm.

SUPERCLASS ACTINOPODA. This superclass encompasses forms with long, thin pseudopods stiffened with microtubules.

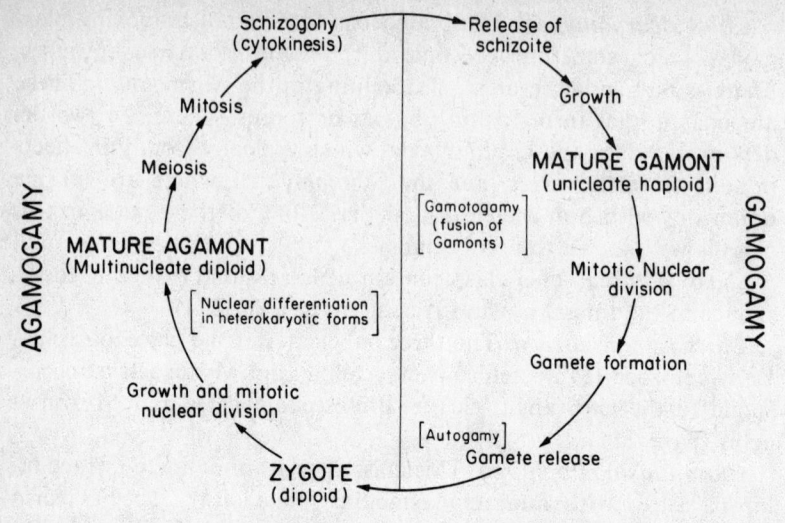

Fig. 13.2. Life cycle of the foraminifera.

Class Acantharea. All the organisms in the Acantharea have a strontium sulfate skeleton composed of 10, 16, or 32 (sometimes more) diametrical or twice as many radial spines, more or less joined in the cell center. The cell is subdivided into an outer cortex (envelope) and a central cell mass by a capsular membrane (inner envelope). All species are planktonic and marine. The class is subdivided on the basis of spine distribution and their behavior in the cell center.

Order Holacanthida. The animals in this order usually have 10 diametrical spines (sometimes 16) which cross in the center. Encystment before sporogenesis is known in some species. Example: *Acanthocolla*.

Order Symphyacanthida. The 20 radial spines of the animals in this order are fused in the cell center or form a small sphere in the cell center by apposition of their basal pyramids. Some of these species are also known to encyst before sporogenesis. Example: *Acantholithium*.

Order Chaunacanthida. This order contains organisms with 20 radial spines that have bases more or less loosely articulated in the cell center. Perhaps all species in this order encyst before sporogenesis. Example: *Gigartacon*.

Order Arthracanthida. These organisms have 20 radial spines

Protozoa

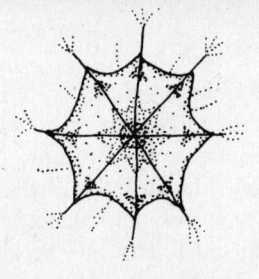

Arcella
∼ X 500

Acantholithium
∼ X 200

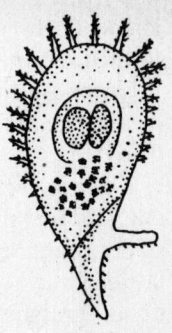

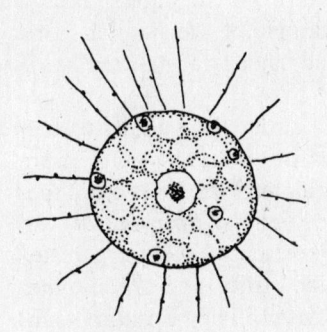

Challengeron
∼ X 200

Thalassiocolla
∼ X 250

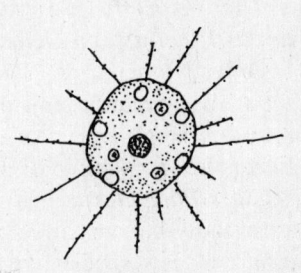

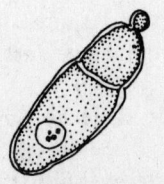

Actinophrys
∼ X 450

Trophozoite of *Gregarina*
∼ X 20

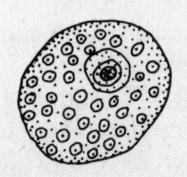

Trophozoite of *Monocystis*
∼ X 300

joined at the cell center by apposition of their bases. No cysts are formed. Example: *Acanthometra*.

Order Actinelida. These animals differ from other orders of the Acantharea in that they have a variable number of radial spines not distributed in the regular way. Example: *Actinelius*.

Class Polycystinea. The skeleton of these organisms is a siliceous one. In the simplest forms, the skeleton is composed of isolated spines, but the advanced forms have solid skeletons of one or more latticed shells with radial spines. The capsular membrane is built up of grossly polygonal plates containing pores. All species are planktonic and marine.

Order Spumellarida. This order contains simple species with one or more isolated spicules and species with complete concentric shells. A capsular membrane with uniformly distributed pores is characteristic of these organisms. An example of the simpler forms is *Thalassiocolla* and examples of the advanced forms are *Coccodiscus lithelius* and *Octodendron*.

Order Nassellarida. The species in this order have a ring or basket-shaped skeleton and capsular membranes with the pores gathered at a single pole. An example of a simpler form in the order is *Eucornis;* an example of a more advanced form is *Botryocella*.

Class Phaeodaria. The spines of the skeleton of these organisms are usually hollow. The skeleton is of mixed silica and organic materials. A few species do not have a skeleton. The capsular membrane is very thick and has one large opening (astropyle) and two smaller openings (parapylae). The Phaeodaria are usually found deeper in the sea than the Polycystiens.

Order Phaeocystida. These organisms lack a skeleton or have a simple cone consisting of free spicules or one with spicules radiating from a common junction. Example: *Aulacantha*.

Order Phaeosphaerida. These forms have large latticed skeletons with wide polygonal meshes. Example: *Cannosphaera*.

Order Phaeocalpida. The skeleton here consists of a small porcelaneous shell with numerous pores and usually one large opening. Example: *Tuscarora*.

Order Phaeogromida. The diatomaceous skeleton has a single, large opening. Example: *Challengeron*.

Order Phaeoconchida. These have a skeleton built of two thick hemispherical valves pressed against each other. Example: *Concharium*.

Order Phaeodendrida. The skeletons of these organisms also are built up of two valves. Unlike those in the previous order, the two valves are not tightly pressed together. Long branching, sometimes latticed spines extend from the surface of the valves. Example: *Coelodendrum.*

Class Heliozoea. These are mainly freshwater actinopods. Only a few species are marine. Their actinopodia radiate on all sides. They do not have a central capsule and many do not have skeletons. The skeletons of those that have them are siliceous or organic. In several orders, the axial fibrils of the axopodia arise from a *central granule* or *centroplast* rather than from a nuclear membrane.

Order Desmothoraci. Desmothoraci do not have a centroplast and are usually enclosed in a spherical latticed organic capsule. Many species are attached by a stalk. Uniflagellated or biflagellated zoospores are produced. Example: *Clathrulina.*

Order Actinophyrida. These organisms do not form skeletons and also do not have a centroplast. The axonemes that stiffen the axopodia of this order are usually visible in the light microscope. Some species have flagella or a flagellated stage. Examples: *Actinophyrs, Actinospherium, Ciliophyrs.*

Order Centrohelida. These have skeletons of siliceous plates and/ or spines; some have organic spicules. The axonemes of the axopodia insert on a centroplast. The nucleus is eccentric. Some species are flagellated or have flagellated stages. Examples: *Acanthocystis* and *Rabdiophrys.*

Order Taxopodia. This order contains a single, very unusual genus of marine actinopod, *Sticholonche.* The characteristics of this organism are such that it could also be considered an acantharian. *Sticholonche* is a bilaterally symmetrical planktonic species that swims by rowing its long, parallel, longitudinal rows of axopodia. The axopodia insert a thick nucleothecia.

PHYLUM LABYRINTHOMORPHA. This marine group has only one class and a single order. The trophic stages form an ectoplasmic network within which glide spindle-shaped or spherical nonamoeboid cells. Mycologists consider this phylum to be an order (or class) Labyrinthulales within the Oomycota.

PHYLUM APICOMPLEXA. Apicomplexa contains a great variety of parasitic species. The unique feature of the phylum, the apical complex, is visualized only with the electron microscope; it consists of a number of smaller parts (polar ring[s], rhoptries, micronemes, conoid, and subpellicular microtubules).

Class Perkinsea. This class has a single order (Perkinsida) and a single genus, *Perkinsus.* The conoid is incomplete in this form.

Class Sporozoea. The Sporozoea includes many well-known genera and species. The conoid, if present, forms a complete cone. Life cycles include both sexual and asexually reproducing stages. Mature organisms move by gliding, body flexion, or undulations. The microgametes of some groups are flagellated. Three subclasses are recognized: Gregarinia, Coccidia, and Piroplasmia.

Subclass Gregarinia. The gregarines are typically extracellular parasites of the digestive tract and body cavities of invertebrates. A few are found in tunicates and Enteropneusta. Mature trophozoites (feeding stage) can be very large (3 to 5 mm). The *mucron* or epimerite, an organelle formed from the conoid, is firmly attached to the host's tissue or embedded in a tissue cell. There are three orders.

Order Archigregarinida. These organisms have a more primitive life cycle with three replicative stages: merogony, gametogony and sporogony. They are parasites of annelids, sipunculids, and ascidians. Example: *Exoschizon.*

Order Eugregarinida. These lack the merogony stage. They are parasites of marine and terrestrial annelids and arthropods. Examples: *Monocystis, Gregarina, Selenidium.* Interestingly, the life cycles of many marine gregarines (e.g., *Gonospora*) are timed to coincide with the spawning of their hosts. In one family, Porosporida, the life cycle involves both a crustacean and a molluscan host. Trophozoites mature in the intestine of a crab or lobster, then produce gymnospores which are released into the sea. Within the gymnospores are merozoites which are released when the gymnospores are eaten by suitable molluscan hosts. The merozoites become differentiated into gamonts which produce gametes. The zygotes formed by fusion of the gametes differentiate into sporozoites which infect a crab or lobster that eats the mollusk.

Order Neogregarinida. The merogony stage is present. These are all insect parasites being found in malpighian tubules, intestine, hemocoel, or fat bodies. Examples: *Gigaductus, Mattesia.*

Subclass Coccidia. The Coccidia are smaller than the gregarines and are usually intracellular parasites. They do not form a mucron or epimerate. The life cycle characteristically includes merogony, gamogony, and sporogony. Most species are vertebrate parasites some of which (e.g., *Plasmodium, Eimeria, Toxoplasma*) cause serious diseases in humans and domestic animals. There are three orders.

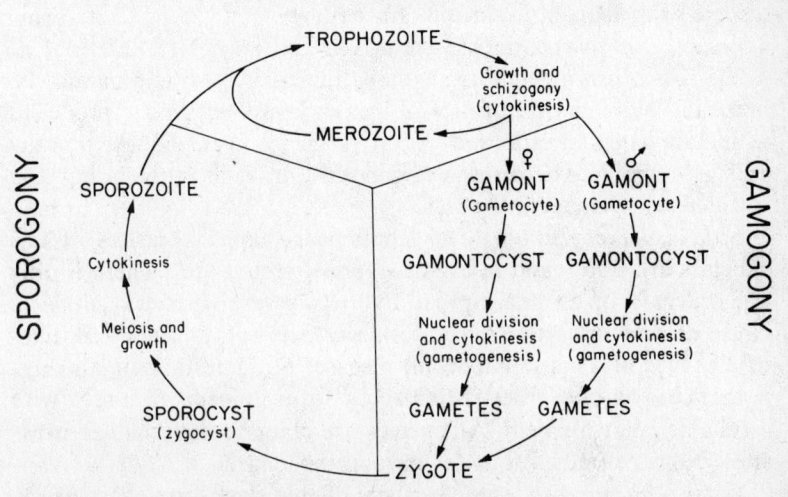

Fig. 13.3. Life cycle of a coccidian (e.g., *Plasmodium*).

Order Agamococcidiida. These forms do not have stages with merogony and gametogony. Example: *Rhytidocystis*.

Order Protococcidiida. These invertebrate parasites lack a merogony stage. Example: *Grellia*.

Order Eucoccidiida. The largest and most important order of Coccidia is the Eucoccidiida. In the suborder in which malaria belongs, the life cycle takes place alternately in vertebrates and in blood-sucking insects (Fig. 13.3). Merogony occurs in endothelial or related tissue cells in the host. The merozoites that are produced can reinvade similar host cells or reinfect erythrocytes. The erythrocytic stages undergo multiple fission (schizogony) to produce more merozoites which reinfect more erythrocytes or which develop into gamonts (gamete-producing cells). If the gamonts are ingested by a mosquito in a blood meal, the gamonts continue their development to produce either a spherical macrogamont (♀) or eight smaller microgametes (♂). The zygote that is formed (ookinete) burrows through the wall of the stomach and attaches on the outside. There it transforms and grows into an oocyst which undergoes multiple fission (sporogony) to produce needle-shaped sporozoites. The sporozoites migrate to the salivary gland where they become ready for inoculation into the next vertebrate host. *Plasmodium falciparum*, the most pathogenic species of malaria which affects humans, probably causes more human deaths in tropical countries than any other disease organism. *Eimeria* spp. are pathogens in the small intestine and ceca of chickens, ducks, geese, and cattle where they are often fatal. *Toxoplasma gondii* is also a pathogen of birds and mammals, including humans. Merozoites of this parasite, which are produced in macrophages, invade and encyst in nerve, liver, kidney, muscle, and lung tissue. The disease is generally mild in humans but can cause severe damage in fetuses.

Subclass Piroplasmia. The small pear-shaped parasites in this subclass are also found in erythrocytes of vertebrates. There is only one order, Piroplasmida. Spread by ticks, several species of *Babesia* cause disease in cattle (red-water fever), horses, and dogs. Species of *Theleria* also cause important diseases of domestic animals.

PHYLUM MICROSPORA. Microspora form unicellular spores with a tubular polar filament. All species are obligate intracellular parasites. None examined thus far have mitochondria.

Class Rudimicrosporea. Members of this class form short, thick, rudimentary polar filaments. Sporulation takes place within a thick-walled sporocyst. The few species that have been described are parasites (hyperparasites) of gregarines which are themselves parasites in annelids. There is but a single order (Metchnikovellida) with only a few genera (*Metchnikovella* and *Amphiacantha*).

Class Microsporea. The spores of the Microsporea contain a thin, coiled, tubular polar filament. There are two orders.

<u>Order Minisporida.</u> These forms have a general tendency toward minimum development of spore organelles. Example: *Hessea*.

<u>Order Microsporidia.</u> These have well-developed spore organelles. *Nosema apis* causes the destructive nosema disease of adult honeybees. *Nosema bobycis* is responsible for the pebrine disease of silkworms. *Nosema* spp. attack fish muscle tissues, producing tumorlike masses. *Ameson michaelis* can cause heavy chalky-grayish infections of the appendages and abdominal muscles of blue crabs. Other species of *Nosema* parasitize helminths, annelids, arthropods, fish, and mammals.

PHYLUM ASCETOSPORA. The organisms in this phylum are characterized by their multicellular spores with one or more sporoplasms. The spores do not contain polar capsules or polar filaments. All species are parasitic.

Class Stellatosporea. These organisms form haplosporosomes. There are two orders.

Order Occulosporida. Spores with more than one sporoplasm form by endogenous buddings. *Marteilia refringens* is a pathogen of European flat oysters and Japanese oysters.

Order Balanosporidia. The species in this order form spores with one sporoplasm and with an orifice covered by an operculum or a diaphragm. Examples: *Haplosporidium, Minchinia, Urosporidium*. *Minchinia nelsoni* has been responsible for mass mortality of oysters on the U.S. mid-Atlantic seaboard.

Class Paramyxea. Bicellular spores consist of a parietal cell and a sporoplasm. There is but a single order, Paramyxida, with a single genus, *Paramyxa*.

PHYLUM MYXOZOA. The parasites in phylum Myxozoa form multicellular spores with one or more polar capsules and sporoplasms and one to three valves.

Class Myxosporea. (Fig. 13.4) Spores of these forms have one or two sporoplasms, and one to six polar capsules (each with a coiled polar filament). There are two orders.

Order Bivalvulida. Spores have two valves. *Myxobolus pfeifferi* causes boil disease of cyprinid fish. Many genera of Myxosporea are parasites of fish, where they develop in the body cavities, gallbladder, urinary bladder, muscles, cartilage, liver, and so forth, reducing their growth, spoiling the commercial value of the flesh, or killing them.

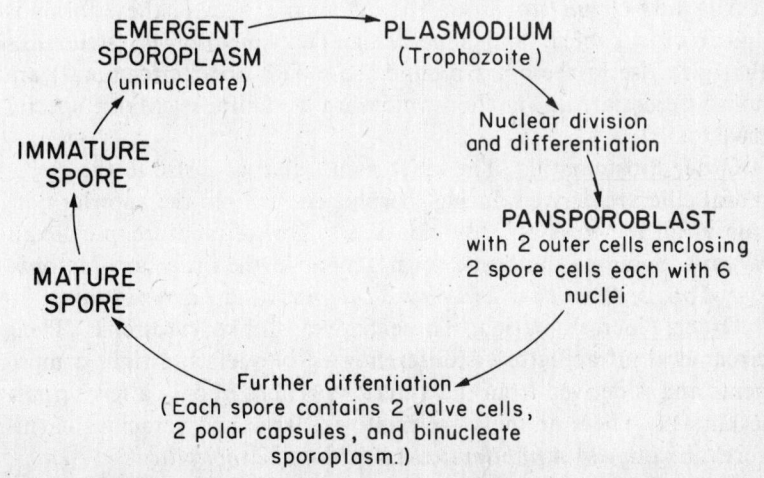

Fig. 13.4. Life cycle of a myxosporean (e.g., *Myxobolus*).

Order Multivalvulida. Spore walls have three or more valves. More seriously pathogenic than the bivalvulids are *Unicapsula, Kudoa,* and *Hexacapsula* which invade the skeletal musculature of different teleosts, causing diseases such as milky or wormy halibut, characteristics that make the fish unmarketable.

Class Actinosporea. Spores have three polar capsules, each enclosing a coiled polar filament, many sporoplasms, and two valves. There is but one order; these animals parasitize the body cavities and gut epithelium of oligochaete worms and sipunculids. *Triactinomyxon ignotum,* for example, parasitizes *Tubifex.*

PHYLUM CILIOPHORA. Phylum Ciliophora is a large one. All members have two types of nuclei (macro- and micronuclei) and cilia and associated subpellicular infraciliature at one or all stages in the life cycle. Many groups divide by transverse binary fission but some reproduce by budding or binary fission. Sexuality involves conjugation, autogamy, or cytogamy. Most species are free-living but some are commensal and others are parasitic.

Class Kinetograminophorea. The oral infraciliature of Class Kinetograminophorea is only slightly distinct from body (somatic) infraciliature. The cytosome is apical, subapical, or midventral. Compound oral or somatic ciliature is not a class feature. Most species are free-living but there are some commensal and parasitic species. There are four subclasses: Gymnostomatia; Vestibuliferia; Hypostomatia; Suctoria.

Subclass Gymnostomatia. The cytostomal area of these forms is apical or subapical and superficial. The kinetosomes (structures that give rise to the cilia) around the mouth area (circumoral) are not differentiated. The body ciliature is uniform. Many species have toxicysts.

Order Prostomatida. The kinetosomes that give rise to the circumoral cilia are derived during morphogenesis from the anterior portion of all the rows of body ciliature (a row of ciliature is called a *kinety*). Some of the best-known genera in the order are *Holophrya, Coleps, Prorodon, Didinium, Dileptus,* and *Lacrymaria.*

Order Pleurostomatida. These have a slitlike cytostome. Their circumoral infraciliature is differentiated into left and right components and is derived from the anterior portions of only a few somatic kineties. These animals are relatively large and voracious carnivores. Examples: *Amphileptus, Litonotus, Loxophyllum.*

Order Primociliatida. The organism *Stephanopogon* is placed in

Unicapsula
∼ X 1,500

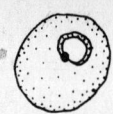

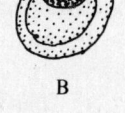

A   B
Ring form Microgametocyte
Plasmodium ovale
in human erythrocyte
∼ X 1,500

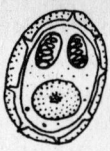

Myxobolus
∼ X 600

Nosema
∼ X 600

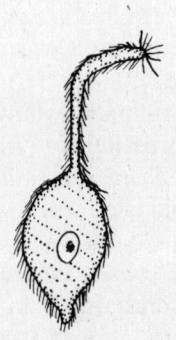

Lacrymaria
∼ X 200

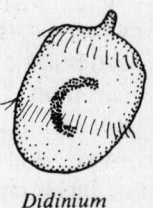

Didinium
∼ X 200

Litonotus
∼ X 200

Loxodes
∼ X 300

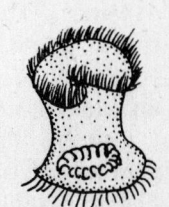

Trichodina
∼ X 300

its own order, Primociliatida. The outstanding characteristic of this unusual marine ciliate is that it has only one type of nucleus with a prominent RNA-rich nucleolus. This is contrasted with the two types of nuceli (macro and micro) found in other ciliates.

Order Karyorelictida. These have a diploid, nondividing macronucleus and an apical or ventral oral area. They are mainly interstitial sand-dwelling forms. Many are carnivorous. Examples: *Geleia, Loxodes, Trachelocerca.*

Subclass Vestibuliferia. As the name implies, the animals in this subclass have a vestibulum or funnel-shaped depressed area at one pole of the body. The vestibulum is lined with cilia and leads directly to a mouth (cytostome-cytopharyngeal complex). Many of the animals are free-living but some are parasites in the digestive tracts of vertebrates or invertebrates.

Order Trichostomatida. These forms do not have reorganized body kineties in their vestibulum. Many species are endocommensal in vertebrate hosts such as humans, cattle, and horses. Examples: *Balantidium, Isotricha, Sonderia, Raabenia.*

Order Entodiniomorphida. These are found in the rumen of cattle, sheep, and other ruminants and the intestine of certain herbivores, including apes. The body ciliature is reduced to unique ciliary tufts or bands. The oral area is sometimes retractable. Internal skeletal plates are present in many species. Examples: *Entodinium, Ophryoscolex.*

Order Copodida. The characteristic that separates the order Copodida from the rest of the Vestibuliferia is a more complex vestibular ciliature. The body of colpodids is often contorted, complicating the steps required for the development of daughter cells. Examples: *Colpoda, Tillina, Woodruffia.*

Subclass Hypostomatia. This subclass is characterized by a ventral cytostome and cyrtos-type (cornucopia-shaped) cytopharyngeal apparatus. The walls of the cytopharynx are strengthened by particular interconnected microtubules and annular sheaths. The development of the mouth area is often very complex, involving the anterior extremities of all or some somatic kineties.

Order Synhymeniida. These animals have an extensive band of perioral ciliature (hypostomial frange) which winds around helically onto part of the dorsal surface of the body, completely, or almost completely, traversing the ventral surface at a level slightly posterior to the oral area. Example: *Orthodonella.*

Order Nassulida. The hypostomial frange on members of the order Nassulida never passes under or to the right of the cytostomal area. It is sometimes reduced to a few "pseudomembranes" located on the left very close to the mouth. Examples: *Nassula, Leptopharynx, Microthorax.*

Order Cyrtophorida. The ciliates in this order have three rows of oral ciliature. The animals are generally flattened and cilia on the body are restricted to the ventral surface. Some species have an adhesive organelle (podite or stylet) at their posterior ends. Cyrtophorids feed mainly on bacteria, but some are histophages and others are ectosymbionts on freshwater and marine fishes and invertebrates. Examples: *Chilodonella, Chlamydodon, Dysteria, Hypocoma.*

Order Chonotrichida. Chonotrichida are vase-shaped, sessile animals that attach by a stalk to the outer surfaces of crustaceans and other invertebrates. Except for the ciliature on the ventral surface, which is displaced to the apical end, the rest of the body is naked. These ciliates reproduce by budding. Examples: *Heliochona, Chonosaurus, Stylochona.*

Order Rhynchodida. These are small forms nearly devoid of somatic ciliature with an apical sucking tube and toxicysts. The sucking tube has a complex structure which is presumably homologous with the cyrtos. These animals are commensals or pathogens on the gills of marine bivalves. Examples: *Gargarius, Ancistrocoma, Sphenophyra.*

Order Apostomatida. This is an order of parasitic ciliates with complex life cycles sometimes involving the alternation of hosts. The cytostome is inconspicuous or absent in certain stages of the life cycle. A glandular complex (rosette) is located near the oral area. The body ciliature is spiralled and often widely spaced. Most species parasitize marine crustacea but some species parasitize freshwater crustaceans, polychaetes, amphipods, isopods, squid, octopus, and even terrestrial mites.

Subclass Suctoria. The adult ciliates in Subclass Suctoria are sessile, generally unciliated, and have many suctorial tentacles armed with haptocysts (missilelike bodies containing lytic enzymes). Asexual reproduction is by budding. Conjugation involves micro- and macroconjugants. These animals are widespread in marine and freshwater habitats. Some species are ectocommensals. There is only one order, Suctorida. Examples: *Ephelota, Podophyra, Acineta, Tokophrya, Heliophrya.*

Class Oligohymenophorea. The characteristics of the oral apparatus separate the ciliates in Class Oligohymenophorea from those in other classes. The oral apparatus is well defined and in a buccal cavity. The oral ciliature is distinct from the somatic ciliature and consists of a membrane (paraoral) on the right and a small number of compound organelles (membranelles, peniculi) on the left of the cytostome.

Subclass Hymenostomatia. The body ciliation of animals belonging to this subclass is uniform and heavy. The buccal cavity is ventral.

Order Hymenostomatida. The buccal cavity of the ciliates in this order is well defined and contains membranelles with infraciliary bases of three to eight rows of kinetosomes. Most of these animals are free-living but some are parasites. There are many well-known genera in the order including *Paramecium, Frontonia, Tetrahymena, Colpodium,* and *Glaucoma. Ichthyophthirius multifiliis* is perhaps the best-known freshwater aquarium parasite, causing the disease known as "white spot" or "itch."

Order Scuticociliatida. The body ciliature here varies from uniform to sparsely ciliated. The buccal ciliature is dominated by a triparte paraoral membrane on the right and several (usually three) membranelles on the left. The most characteristic feature of the group is the pattern of development of the oral region which involves the appearance of a J-shaped scutica at one step in the process. The order is particularly abundant in brackish or marine habitats and as endocommensals of sea urchins, mollusks, polychaetes, and oligochaetes.

Order Astomatida. As the name implies, the ciliates in the order Astomatida lack a mouth. They are all endoparasites. Most are found in oligochaetes, but they are also found in other annelids, tailed amphibians, mollusks, and turbellarians.

Subclass Peritrichida. These forms are quite distinctive, with goblet-shaped or disk-shaped bodies. The ciliature on the body is reduced, but the entire apical end of the animal is encircled by a prominent oral ciliary field which dips into a funnel-shaped buccal cavity (infundibulum). There is only one order, Peritrichida. Many of the sessile genera are quite familiar (e.g., *Vorticella, Epistylis, Carchesium, Zoothamnium*). Some less familiar genera (e.g., *Trichodina, Urceolaria*) have complex organelles for attachment to freshwater and marine hosts. A disease known as trichodiniasis is caused by species of *Trichodina* which parasitize the gills of fish.

Colpoda
~ X 300

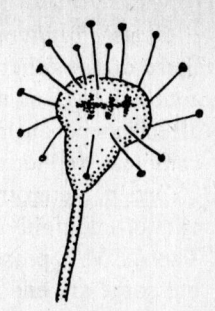

Vorticella
~ X 300

Ephelota
~ X 350

Paramecium
~ X 200

Euplotes
~ X 200

Class Polyhymenophorea. Members of this class differ from those of the other two by their well-developed, conspicuous, adoral zone (AZM) of numerous buccal or peristomial organelles. These often extend out onto the body surface. The body ciliature may be complete, reduced, or specialized into cirri.

Order Heterotrichida. These animals are generally very large and often highly contractile. The macronucleus is often beaded. The AZM is proportionally quite large. Most species are free-living but some are endoparasites or endocommensals in insects and other arthropods, mollusks, oligochaetes, and lower vertebrates. Many of the ciliates in the order are very well known (e.g., *Blepharisma, Metopus, Spirostomum, Stentor, Nyctotherus, Folliculina*).

Order Odontostomatida. The relatively small (~ 50 μm) ciliates in this order are laterally compressed and wedge-shaped. They have a rigid armorlike pellicle and often have posterior spines. The body ciliature is reduced to eight or nine membranelles and no paraoral membrane. These animals occur primarily in putrefying organic freshwaters, feeding preferentially on sulfur bacteria. Examples: *Saprodinium, Mylestoma.*

Order Oligotrichida. The distinguishing features of this order are an extensive AZM often divided into two parts, one within the buccal cavity and one out onto the anterior body surface; a ciliary membrane lying along the right border of the buccal cavity (paraoral membrane) which arises out of a distinctive kinety of nonzigzag kinetosomes; and body ciliature often reduced to a few widely spaced shortened rows of specialized cilia or bristles. Animals in one suborder, the Tintinnina, have a lorica. Examples: *Halteria, Strobilidium, Strombidium, Codonella, Tintinnopsis.*

Order Hypotrichida. These ciliates are flattened dorsoventrally and have bundles of fused cilia (cirri) on their ventral surfaces and rows of widely spaced "sensory-bristle" cilia on their dorsal surfaces. There are many species widely distributed in freshwater and marine habitats. Examples: *Euplotes, Aspidisca, Oxytricha, Stylonychia, Holosticha, Kerona, Urostyla.*

SUPPLEMENTARY READING

1. INTRODUCTION TO MICROBIOLOGY

Ainsworth, G. C., and Sneath, P. H. A., eds. 1962. *Microbial classification.* New York: Cambridge Univ. Press.

Barghoorn, E. S. 1971. The oldest fossils. *Scientific American* 22(5):30–42 (May).

Brock, T. D., ed. and translator. 1975. *Milestones in microbiology.* Washington, D. C.: American Society for Microbiology.

Bulloch, W. 1938. *The history of bacteriology.* London: Oxford Univ. Press.

DeKruif, P. 1926. *Microbe hunters.* New York: Harcourt, Brace.

Dobell, C. 1932 (reprint 1960). *Anthony von Leeuwenhoek and his little animals.* New York: Dover.

Dubos, R. J. 1950 (reprint 1976). *Louis Pasteur: free lance of science.* New York: Scribner.

Groves, D. I., Dunlap, J. S. R., and Buick, R. 1981. An early habitat of life. *Scientific American* 245 (4):64–73 (Oct.).

Lechavalier, H. A., and Solotorovsky, M. 1965. *Three centuries of microbiology.* New York: McGraw-Hill.

Leedale, G. F. 1974. How many are the kingdoms of organisms? *Taxon* 23:261–270.

Margulis, L. 1974. Five kingdom classification and the origin and evolution of cells. In *Evolutionary biology I,* eds. M. Hecht, W. Steerer, and T. Dobzhansky. New York: Plenum.

Margulis, L. 1974. The classification and evolution of prokaryotes and eukaryotes. In *Handbook of genetics,* ed. R. C. King. New York: Plenum.

Margulis, L. 1975. The microbes' contribution to evolution. *Biosystems* 7:266–292.

Margulis, L. 1981. Symbiosis in cell evolution: Life and its environment on the early earth. San Francisco: Freeman.

Swain, F. M. 1969. Paleomicrobiology. *Annual Review of Microbiology* 23:455.

Watson, J. D. 1968. *The double helix.* New York: Atheneum.
Whittaker, R. H. 1969. New concepts of kingdoms of organisms. *Science* 163:150–160.

2. TOOLS AND METHODS OF MICROBIOLOGISTS

Barer, R. 1974. Microscopes, microscopy and microbiology. *Annual Review of Microbiology* 28:371–389.
Bradbury, S. 1968. *The microscope, past and present.* New York: Pergamon Press.
Everhart, T. E., and Hayes, T. L. 1972. The scanning electron microscope. *Scientific American* 227(3):43 (July).
Lawrence, C. A., and Block, S. S. 1968. *Disinfection, sterilization and preservation.* Philadelphia: Lea & Febiger.
Lennette, E. H., Spaulding, E. H., and Truant, J. P., eds. 1974. *Manual of clinical microbiology.* 2nd ed. Washington D. C.: American Society for Microbiology.
Phillips, G. B., and Miller, W. S., eds. 1972. *Industrial sterilization.* Durham, N. C.: Duke Univ. Press.

3. PROCARYOTIC CELL STRUCTURE

Aronson, A. I., and Fitz-James, P. 1976. Structure and morphogenesis of the bacterial spore coat. *Bacteriological Reviews* 40:360–402.
Costerton, J. W., Ingram, J. M., and Cheng, K. J. 1974. Structure and function of the cell envelope of gram-negative bacteria. *Bacteriological Reviews* 38:87–110.
Ottow, J. C. G. 1975. Ecology, physiology and genetics of fimbriae and pili. *Annual Review of Microbiology* 30:451–482.
Pelczar, M. J. (chairman of committee). 1975. *Manual of microbiological methods.* New York: McGraw-Hill.
Sharon, N. 1969. The bacterial cell wall. *Scientific American* 220(5):92–98 (May).
Silverman, M., and Simon, M. I. 1977. Bacterial flagella. *Annual Review of Microbiology* 31:397–420.
Stanier, R. Y., Rogers, H. J., and Ward, B. J., eds. 1978. Relations between structure and function in the prokaryotic cell. In *Society for general microbiology, symposium 28.* London: Cambridge Univ. Press.
Stoeckenius, W. 1976. The purple membrane of salt-loving bacteria. *Scientific American* 234(6):38–46. (June).

4. MICROBIAL GROWTH

Alexander, M. 1971. Biochemical ecology of microorganisms. *Annual Review of Microbiology* 25:361–392.

Brock, T. D. 1967. Life at high temperatures. *Science* 158:1012–1019.
Brock, T. D. 1972. Microbial growth rates in nature. *Bacteriological Reviews* 35:39.
Dawson, P. S. S., ed. 1975. Microbial growth. In *Benchmark papers in microbiology*. Stroudsburg, Pa.: Dowden, Hutchinson and Ross.
Jannasch, H. W., and Wirsen, C. O. 1977. Microbial life in the deep sea. *Scientific American* 236(6):42–52 (June).
Kubitschek, H. E. 1970. Introduction to research with continuous cultures. Englewood Cliffs, N. J.: Prentice-Hall.
Meadow, P. M. and Pirt, S. J., eds. 1960. Microbial growth. In *Society for general microbiology, symposium 19*. New York: Cambridge Univ. Press.
Meynell, G. G., and Meynell, E. 1965. Theory and practice in experimental bacteriology. New York: Cambridge Univ. Press.
Monod, J. 1949. The growth of microbial cultures. *Annual Review of Microbiology* 3:371–427.
Morita, R. Y. 1975. Psychrophilic bacteria. *Bacteriological Reviews* 39:114–167.

5. MICROBIAL NUTRITION AND METABOLISM

Kelly, D. P. 1971. Autotrophy: concepts of lithotrophic bacteria and their organic metabolism. *Annual Review of Microbiology* 26:117–210.
Lehninger, A. L. 1975. *Biochemistry*. 2nd ed. New York: Worth.
Mandelstram, J. and McQuellen, M. 1973. *The biochemistry of bacterial growth*. 2nd ed. New York: Wiley.
Moat, A. G. 1979. *Microbial physiology*. New York: Wiley-Interscience.
Philips, D. C. 1966. The three-dimensional structure of an enzyme molecule. *Scientific American* 215(5):78 (Nov.).
Rich, A., and Kim, S. H. 1978. The three-dimensional structure of transfer RNA. *Scientific American* 238(1):52–62 (Jan.).
Smith, A. J., and Hoares, D. S. 1977. Specialist phototrophs, lithotrophs and methyltrophs: A unity among a diversity of prokaryotes? *Bacteriological Reviews* 41:419–448.
Watson, J. D. 1976. *Molecular biology of the gene*. 3rd ed. Menlo Park, Calif.: W. A. Benjamin.

6. VIRUSES

Baltimore, D. 1971. Expression of animal virus genomes. *Bacteriological Reviews* 35:235–241.
Baltimore, D. 1976. Viruses, polymerases and cancer. *Science* 192:632–636.

Cairns, J. 1975. The cancer problem. *Scientific American* 233(5):64–78 (Nov.).

Calender, R. 1970. Regulation of phage development. *Annual Review of Microbiology* 24:241–296.

Campbell, A. M. 1976. How viruses insert their DNA into the DNA of the host cell. *Scientific American* 235(6):103–113 (Dec.).

Diener, T. O. 1981. Vivoids. *Scientific American* 244(1):66–73 (Jan.).

Epstein, M. A. 1978. Epstein-Barr virus as the cause of a human cancer. *Nature* 274:740.

Fenner, F. 1974. The classification of viruses: why, when and how. *Australian Journal of Experimental Biology and Medical Science* 52:223–240.

Fenner, F., and White, D. O. 1976. *Medical virology*. New York: Academic Press.

Fiddes, J. C. 1977. The nucleotide sequence of viral DNA. *Scientific American* 237(6):54–67 (Dec.).

Gajdusek, D. C. 1977. Unconventional viruses and the origin and disappearance of Kuru. *Science* 197:943–960.

Holland, J. J. 1974. Slow, inapparent and recurrent viruses. *Scientific American* 230(2):32–40 (Feb.).

Knight, C. A. 1974. Molecular virology. New York: McGraw-Hill.

Luria, S. E., and Darnell, J. E. 1967. *General virology*. New York: Wiley.

Padan, E., and Shilo, M. 1973. Cyanophage viruses attacking blue-green algae. *Bacteriological Reviews* 37:343–370.

Rafferty, K. A. 1973. Herpes, viruses, and cancer. *Scientific American* 229(4):26–33 (Oct.).

Spector, D. H., and Baltimore, D. 1975. The molecular biology of virus. *Scientific American* 232(5):24–31 (May).

Suskind, M. M., and Botstein, D. 1978. Molecular genetics of bacteriophage P 22. *Microbiological Reviews*, 32:385–413.

Temin, H. M. 1972. RNA directed DNA synthesis. *Scientific American* 226(1):24–33 (Jan.).

Wagner, E. K. 1974. The replication of herpes virus. *American Scientist*, 62:584–593.

7. GENETICS OF MICROORGANISMS

Bachmann, B. J., Low, K. B., and Taylor, A. L. 1976. Recalibrated linkage map of E. coli. K-12. *Bacteriological Reviews* 116–167.

Brown, D. 1973. The isolation of genes. *Scientific American* 229(2):20–29 (Aug.).

Clowes, R. 1973. The molecule of infectious drug resistance. *Scientific American* 223(4):18–27 (Apr.)

Cohen, S. N. 1975. The manipulation of genes. *Scientific American* 233(1):24–33 (July).

Cohen, S. N. 1976. Transposable genetic elements and plasmid evolution. *Nature* 263:231–238.

Curtiss, R. 1976. Genetic manipulation of microorganisms: potential benefits and biohazards. *Annual Review of Microbiology* 30:507–533.

Fincham, J. R., and Day, P. 1971. Fungal genetics. 3rd ed. Philadelphia: Davis.

Fogel, S., and Mortimer, R. K. 1971. Recombination in yeast. *Annual Review of Genetics* 8:347.

Fox, M. S. 1978. Some features of genetic recombination in procaryotes. *Annual Review of Genetics* 12:47–68.

Guarente, L., Roberts, T. M., and Ptashne, M. 1980. A technique for expressing eukaryotic genes in bacteria. *Science* 209:1428–1430.

Hopwood, D. A. 1973. Advances in *Streptomyces coelicolor* genetics. *Bacteriological Review* 37:371.

Low, K. B., and Porter, D. D. 1978. Modes of gene transfer and recombination in bacteria. *Annual Review of Genetics* 12:249–287.

Meynell, G. G. 1973. *Bacterial plasmids*. Cambridge, Mass.: M.I.T. Univ. Press.

Mulligan, R. C., and Berg, P. 1980. Expression of a bacterial gene in mammalian cells. *Science* 209:1422–1427.

Peden, K. W. C., Pipas, J. M., Pearson-White, S., and Nathans, D. 1980. Isolation of mutants of an animal virus in bacteria. *Science* 209:1392–1396.

Roth, J. R. 1974. Frameshift mutations. *Annual Review of Genetics* 8:213.

Sager, R. 1972. Cytoplasmic genes and organelles. New York: Academic Press.

Sanderson, K. E., and Hartman, P. E. 1978. Linkage map of *Salmonella typhimurium*. *Microbiological Reviews* 42:471–519.

Watson, J. D. 1976. Molecular biology of the gene. 3rd ed. Menlo Park, Calif.: W. A. Benjamin.

8. SYMBIOSIS

Ahmadjian, V. 1970. The lichen symbiosis: its origin and evolution. In *Evolutionary biology*. vol. 4. eds. T. Dobzhansky, M. K. Hecht, and W. C. Streere. New York: Appleton-Century-Crofts.

Brill, W. J. 1977. Biological nitrogen fixation. *Scientific American* 236(3):68–81 (Mar.).

Buchner, P. 1965. *Endosymbiosis of animals with plant microorganisms*. New York: Wiley.

Hungate, R. E. 1966. *The rumen and its microbes*. New York: Academic Press.

Hungate, R. E. 1975. The rumen microbial ecosystem. *Annual Review of Ecology and Systematics* 6:39–66.

Lewis, D., and Swan, H. 1971. The role of intestinal flora in animal nutrition. In *Microbes and biological productivity, Society for general microbiology symposium 21*, eds. D. E. Hughes and A. H. Rose. New York: Cambridge Univ. Press.

Muscatine, L., and Greene, R. W. 1973. Chloroplasts and algae as symbionts in molluscs. *International Review of Cytology* 56:137.

Quispel, A. 1974. *The biology of nitrogen fixation.* New York: Elsevier.

Smith, D. C. 1962. The biology of lichen thalli. *Biological Reviews* 37:537–570.

Smith, D., Muscatine, L., and Lewis, O. 1969. Carbohydrate movement from autotrophs to heterotrophs in parasitic and mutualistic symbiosis. *Biological Reviews* 44:17.

Stewart, W. D. P., ed. 1976. *Nitrogen fixation of free-living microorganisms.* New York: Cambridge Univ. Press.

Taylor, D. L. 1973. Algal symbiosis of invertebrates. *Annual Review of Microbiology* 27:171–187.

9. MICROBIAL ACTIVITIES IN THE BIOSPHERE

Brill, W. J. 1977. Biological nitrogen fixation. *Scientific American* 236(3):68–81 (Mar.).

Brock, T. D. 1966. *Principles of microbial ecology.* Englewood Cliffs, N.J.: Prentice-Hall.

Delwiche, C. C. 1970. The nitrogen cycle. *Scientific American* 223(3):136–146 (Sept.).

Jannasch, H. W. 1970. Microbial turnover of organic matter in the deep sea. *Bioscience* 29:228–232.

Kellogg, W. W., Cadie, R. D., Allen, E. R., Lazarus, A. L., and Martell, E. A. 1972. The sulfur cycle. *Science* 175:587–596.

Wirsen, C. O., and Jannasch, H. W. 1976. Decomposition of solid organic materials in the deep sea. *Environmental Science Technology* 10:880–886.

10. APPLIED AND INDUSTRIAL MICROBIOLOGY

Aharonowitz, Y. and Cohen, G. 1981. The microbiological production of pharmaceuticals. *Scientific American* 245(3):140–152 (Sept.).

Amerine, M. A. 1974. Wine. *Scientific American* (Aug.).

Demain, A. L. 1981. Industrial microbiology. *Science* 214:987–995.

Demain, A. L., and Solomon, N. A. 1981. Industrial microbiology. *Scientific American,* 245(3):66–75 (Sept.).

Eveleigh, D. E. 1981. The microbiological production of industrial chemicals. *Scientific American* 245(3):180–196 (Sept.).

Foster, E. M., Nelson, F. E., Speck, M. L., Doetsch, R. N., and Olson, J. C. 1957. *Dairy microbiology.* Englewood Cliffs, N. J.: Prentice-Hall.

Hawkes, H. A. 1961. *The ecology of waste water treatment.* New York: Pergamon Press.

Hopwood, D. A. 1981. The genetic programming of industrial micoorganisms. *Scientific American* 245(3):91–102 (Sept.).

Kaplan, A. M. 1971. Industrial microbiology: concepts, challenges and motivation. *Bioscience* 21:468.

Kleyen, J., and Hough, J. 1971. The microbiology of brewing. *Annual Review of Microbiology* 25:586–608.

Nickerson, J. T., and Sinskey, A. J. 1972. *Microbiology of foods and food processing.* New York: Elsevier.

Phaff, H. J. 1981. Industrial microorganisms. *Scientific American* 245(3):76–89 (Sept.).

Porter, J. R. 1974. Microbiology and the food and energy crises. *American Society of Microbiologists News* 40:813–832.

Rose, A. H. 1981. The microbiological production of food and drink. *Scientific American* 245(3):126–138 (Sept.).

Ryther, J. H., and Goldman, J. C. 1975. Microbes as food in mariculture. *Annual Review of Microbiology* 29:429–443.

Taber, W. A. 1976. Wastewater microbiology. *Annual Review of Microbiology* 30:263–277.

11. MICROBIAL PATHOGENS AND DISEASES OF HUMANS

Baglioni, C., and Nilsen, T. W. 1981. The action of interferon at the molecular level. *American Scientist* 69:392–399.

Burke, D. C. 1977. The status of interferon. *Scientific American* 236:42–62 (Apr.).

Burnet, MacF., and White, D. B. 1972. *Natural history of infectious diseases.* 4th ed. London: Cambridge Univ. Press.

Capra, J. D., and Edmundson, A. B. 1977. The antibody combining site. *Scientific American* 236:50–59 (Jan.).

Cooper, M. D., and Lawton, A. R. 1974. The development of the immune system. *Scientific American* 231:58–72 (Nov.).

Cunningham, B. A. 1977. The structure and function of histocompatibility antigens. *Scientific American,* 237:96–107 (Oct.).

Doolittle, R. F. 1981. Fibrinogen and fibrin. *Scientific American* 245(6):126–135 (Dec.).

Edelman, G. M. 1970. The structure and function of antibodies. *Scientific American* 223:34–53 (Aug.).

Fenner, F., and White, D. D. 1976. *Medical virology.* 2nd ed. New York: Academic Press.

Freimer, E. H., and McCarty, M. 1965. Rheumatic fever. *Scientific American* 213:66–75 (Dec.).

Greenberg, B. 1965. Flies and disease. *Scientific American* 213:92–99 (July).

Hawking, F. 1970. The clock of the malaria parasite. *Scientific American* 222:123–131. (June).

Henderson, D. A. 1977. The eradication of smallpox. *Scientific American* 235:25–33 (Oct.).

Henle, W. E., Henle, G., and Lennette, E. T. 1979. The Epstein-Barr virus. *Scientific American* 241:48–59 (July).

Hirschorn, N., and Greenough, W. B. 1971. Cholera. *Scientific American* 225:15–21 (Aug.).

Jerne, N. K. 1973. The immune system. *Scientific American* 229:52–61 (July).

Kabat, E. A. 1976. *Structural concepts in immunology and immunochemistry.* 2nd ed. New York: Holt, Rinehart and Winston.

Kadis, S., Monti, T. C., and Aji, S. J. 1969. Plaque toxin. *Scientific American* 220:92–103 (Mar.).

Kaplan, M. N., and Webster, R. G. 1977. The epidemiology of influenza. *Scientific American* 237:88–106 (Dec.).

Langer, W. L. 1964. The black death. *Scientific American* 210:114–121 (Feb.).

Langer, W. L. 1976. The prevention of smallpox before Jenner. *Scientific American* 234:112–117 (Jan.).

Meyer, M. M. 1973. The complement system. *Scientific American* 229:54–66 (Nov.).

Melnick, J. L., Pressman, G. R., and Hollinger, F. B. 1977. Viral hepatitis. *Scientific American* 237:44–65 (July).

Raff, M. C. 1976. Cell surface immunology. *Scientific American* 234:30–39 (May).

Rose, N. R., and Friedman, H. 1976. *Manual of clinical immunology.* Washington, D. C.: American Society for Microbiology.

Rosebury, T. 1971. *Microbes and morals.* New York: Viking Press.

Werner, G. H., Latte, B., and Contini, A. 1964. Trachoma. *Scientific American* 210:79–87 (Jan.).

Zinsser, H. 1935. Rats, lice and history. Boston: Little, Brown.

12. SUMMARY OF PROCARYOTIC GROUPS

Barksdale, L. and Kim, K. S. 1977. Mycobacterium. *Bacteriological Reviews* 41:217–372.

Baumann, P. and Baumann, L. 1977. Biology of the marine enterobacteria: *Beneckea* and *Photobacterium. Annual Review of Microbiology* 31:39–61.

Supplementary Reading

Bovarnick, M. 1955. Rickettsiae. *Scientific American* 192:74–79 (Jan.).

Buchanan, R. E., and Gibbons, N. E. coeds. with an editorial board and contributions of 128 colleagues. 1974. *Bergey's manual of determinative bacteriology.* Baltimore: Williams and Wilkins.

Becker, Y. 1978. The Chlamydia: molecular biology of procaryotic obligate parasites of eucaryocytes. *Microbiological Reviews* 42:274–306.

Belser, L. W. 1979. Population ecology of nitrifying bacteria. *Annual Review of Microbiology* 33:309–334.

Canale-Parola, E. 1977. Physiology and evolution of spirochetes. *Bacteriological Reviews* 41:181–204.

Canale-Parola, E. 1978. Motility and chemotaxis of spirochetes. *Annual Review of Microbiology* 32:69–100.

Clayton, R. K., and Sistrom, W. R. eds. 1978. The photosynthetic bacteria. New York. Plenum Press.

Colby, J., Dalton, H., and Whittenbury, R. 1979. Biological and biochemical aspects of microbial growth on C_1 compounds. *Annual Review of Microbiology* 33:481–518.

Ensign, J. C. 1978. Formation, properties, and germination of Actinomycete spores. *Annual Review of Microbiology* 32:185–220.

Goodfellow, M., and Minnikin, D. E. 1977. Nocardioform bacteria. *Annual Review of Microbiology* 31:159–180.

Hastings, J. W., and Nealson, K. H. 1977. Bacterial bioluminescence. *Annual Review of Microbiology* 31:549–595.

Holt, S. C. 1978. Anatomy and chemistry of spirochetes. *Microbiological Reviews* 42:114–160.

Kaiser, D., Manoil, C., and Dworkin, M. 1979. Myxobacteria: cell interactions, genetics and development. *Annual Review of Microbiology* 33:595–640.

Krieg, N. R. 1976. Biology of the chemoheterotrophic spirilla. *Bacteriological Reviews* 40:55–115.

Krulwick, T. A., and Pelliccione, N. J. 1979. Catabolic pathways of coryneforms, nocardias, and mycobacteria. *Annual Review of Microbiology* 33:95–112.

Mah, R. A., Ward, D. M., Baresi, L., and Glass, T. L. 1977. Biogenesis of methane. *Annual Review of Microbiology* 31:309–341.

Nienhaus, F. and Sikora, R. A. 1979. Mycoplasmas, spiroplasmas and Rickettsia-like organisms as plant pathogens. *Annual Review of Phytopathology* 17:37–58.

Parenti, F., and Caronelli, C. 1979. Members of the genus *Actinoplanes* and their antibiotics. *Annual Review of Microbiology* 33:389–413.

Pfennig, N. 1977. Phototrophic green and purple bacteria: a comparative review. *Annual Review of Microbiology* 31:275–290.

Quayle, J. R. 1972. The metabolism of one-carbon compounds by microorganisms. *Advances in Microbial Physiology* 7:119–203.

Razin, S. 1978. The mycoplasmas. *Microbiological Reviews* 42:414–470.
Rippka, R., Dervelles, J., Waterbury, J., Herdman, M., and Stanier, R. 1979. Generic assignments, strain histories and properties of pure cultures of cyanobacteria. *Journal of General Microbiology* 111:1–61.
Schmidt, E. L. 1979. Initiation of plant root-microbe interactions. *Annual Review of Microbiology* 33:355–376.
Shapiro, L. 1976. Differentiation in the *Caulobacter* cell cycle. *Annual Review of Microbiology* 30:377–408.
Thauer, R. K., Jungermann, K., and Decker, K. 1977. Energy conservation in chemotrophic anaerobic bacteria. *Bacteriological Reviews* 41:100–180.
VanVeen, W. L., Mulder, E. G., and Deinema, M. H. 1978. The *Sphaerotilus-Leptothrix* group of bacteria. *Microbiological Reviews* 42:329–356.
Walsby, A. E. 1977. The gas vacuoles of blue-green algae. *Scientific American* 237:90–97 (Aug.).
Whittenbury, R., and Dow, C. S. 1977. Morphogenesis and differentiation in *Rhodomicrobium varniellii* and other budding and prosthecate bacteria. *Bacteriological Reviews* 41:754–808.
Zeikus, J. G. 1977. The biology of methanogenic bacteria. *Bacteriological Reviews* 41:514–541.

13. EUCARYOTIC MICROORGANISMS

Alexopoulos, C. J., and Bold, H. C. 1967. *Algae and fungi.* New York: Macmillan.
Alexopoulos, C. J., and Mims, C. W. 1979. *Introductory mycology.* 3rd ed. New York: Wiley.
Bold, H. C., and Wynne, M. J. 1978. Introduction to the algae. Englewood Cliffs, N. J.: Prentice-Hall.
Chapman, V. J. 1968. *The algae.* New York: Macmillan.
Corliss, J. O. 1979. *The cilicated protozoa.* 2nd ed. Oxford: Pergamon Press.
Curtis, H. 1968. *The marvelous animals. An introduction to the protozoa.* Garden City, New York: Natural History Press.
Grell, K. G. 1973. *Protozoology.* Berlin: Springer-Verlag.
Jahn, T. L., Bovee, E., and Jahn, F. F. 1979. *How to know the protozoa.* Dubuque, Iowa: Brown.
James, D. E. 1978. *Culturing algae.* Carolina Biological Supply Co., Burlington, N. C.
Levine, N. D., and 16 colleagues. 1980. A new revised classification of the protozoa. *Journal of Protozoology* 27:37–58.
Poindexter, J. S. 1971. *Microbiology, and introduction to protists.* New York: Macmillan.

Prescott, G. W. 1968. *The algae: a review.* Boston: Houghton Mifflin.
Sleigh, M. 1973. *The biology of protozoa.* London: Edward Arnold.
Vickerman, K., and Cox, F. E. G. 1967. *The protozoa.* Boston: Houghton Mifflin.

INDEX

Acantharea, 238
Acanthocolla, 238
Acanthocystis, 241
Acantholithium, 238, 239
Acanthometra, 240
Acarpomyxea, 230, 237
Acetabularia, 210, 211
Acetobacter, 134, 188
N-Acetylglucosamine, 40
N-Acetylmuramic acid, 40
Acholeplasma, 205
Achromobacteria, 130
Acid-fast stain, 19
Acineta, 249
Acrasea (Acrasiomycota), 6, 220, 223, 230, 237
Actinelida, 240
Actinelius, 240
Actinobacillus, 188
Actinomyces (Actinomycetales), 199–200
Actinophyrida, 241
Actinophyrs, 239, 241
Actinoplanaceae, 200–201
Actinopoda, 237
Actinospherium, 241
Actinosporea, 246
Active site, 62
Adansonian taxonomy, 9
Adenosine triphosphate, *see* ATP entries
Aeciospores, 222, 229
Aegyptianellosis, 203
Aerobes, 58
Aeromonas, 138, 189
Aerophobes, 58
Aerotolerant anaerobes, 58
Aflatoxin, 145, 227
African tick typhus, 156
Agamococcidiida, 243
Agglutination, 166
Agnotobiotic mixture, 26
Agrobacterium, 185
Ajellomyces dermatitidis, 227
Akinetes, 208, 210

Alcaligens, 188
Algae, 206–220
  correspondence to protozoan groups, 232
  endosymbiotic, 117–118, 208
Alimentary toxic aleukia, 145–146
Alkaline return, 30, 53
Allogromia, 237
Allomyces, 223
Allophycocyanin B, 175
Almond leaf curl disease, 227
Alpha toxin, 154
Amblyomma, 202
Ameson michaelis, 244
Amino acid synthesis, 83, 89, 135–136
Ammonia assimilation, 79–80
Amoeba, 235, 236
Amoebic dysentery, 144, 235
Amoebic meningoencephalitis, 144
AMP, cyclic, 143
Amphiacantha, 244
Amphibolic pathways, 78
Amphiesma, 218
Amphileptus, 246
Anabaena, 124, 176, 177
Anaerobes, 58
Anaerobic respiration, 76–77
Anamnestic reaction, 167
Anaphylactic antibody, 164
Anaplasmataceae, 202, 203
Anaplerotic reactions, 78
Ancistrocoma, 249
Animaliae, 5
Anopheles mosquito, 157–158
Antheridium, 222
Anthrax, 13, 147
Antibiotics, 13, 137, 201, 227
Antibodies, 163–164
Antibody production, 167–168
Anticodon, 85
Antigenic determinants, 163
Antigens, 163
Antiserum, 165
Antitoxin, 165
Aphanomyces, 224

266 Index

Apicomplexa, 231, 241
Aplanospores, 208, 221
Apoenzyme, 63
Apostomatida, 249
Aquaspirillum, 182
Arachnia propionica, 200
Arboviruses, 157
Arbuscules, 116
Arcellinida, 235, 239
Archigregarinida, 242
Arthacanthida, 238
Arthrobacter, 198
Arthroderma, 227
Arthrospores, 221
Ascetospora, 231, 244
Ascocarp, 222
Ascomycota, 6, 221, 226–227
Ascophyllum, 212
Ascospores, 112, 222
Aseptic meningitis, 143, 144
Aseptic technique, 27
Ashbya gossypii, 136
Asian tick typhus, 156
Aspartate family, 89
Aspergilloses, 154, 227
Aspergillus, 113, 129, 137, 145, 154, 227
Aspidisca, 252
Assimilation, 78–82
  ammonia, 79–80
  nitrate and nitrite, 79
  nitrogen, 80–81
  phosphorus, 82
  sulfur, 81
Assimilatory nitrate reduction, 125
Astasia, 212
Asticcacaulis, 179, 180
Astomatida, 250
Asymmetric mode replication, 98
Athiorhodaceae, 64
Athlete's foot, 153
Atopic antibody, 164
ATP (adenosine triphosphate), 36, 60–61, 66, 75, 117
Aulacantha, 240
Autoclave, 30
Autotrophs, 60
Auxospore, 216
Auxotrophic mutants, 78
Avery, O. T., 14, 103, 107
Avian spirochetosis, 181–182
Axenic cultures, 26, 28, 115, 171
Azomonas, 185
Azotobacter, 124, 183, 184–185
Azotobacteraceae, 184–185
Azotococcus, 185

Babesia, 244
Bacillaceae, 195
Bacillariophyta, 6, 212, 216–218
Bacillus, 38
Bacillus, 124, 130, 136, 137, 141, 192, 195
Bacitracin, 137
Bacteria
  morphology of, 37–38
  recombination in, 14, 104, 106–108
  size of, 37
  transformation in, 14
Bacterial capsules, 141
Bacterial dysentery, 142–143, 188
Bacteriophages, 93, 96
Bacteriorhodopsin, 186
Bacteroidaceae, 189–190
Bacteroides, 118, 138, 189, 190
Baker's cheese, 132
Baker's yeast, 227
Balanospiridia, 245
Balantidiasis, 144
Balantidium, 144, 248
Baltimore, D., 102
Barbulanympha, 234
Barrier filter, 21
Bartonellaceae, 202–203
Basidiocarps, 222
Basidiomycota, 6, 222, 228–230
Basidiospores, 222, 229, 230
Basidium, 222, 228
B cells, 167–168
Bdellovibrio, 180, 182, 184
Beadle, George, 14, 103
Beer, 133
Beggiatoa, 178
Beijerinck, M., 14
Beijerinckia, 124, 183, 185
Beneckea, 189
Benzer, 105
Bergey's Manual of Determinative Bacteriology, 169
Berkeleya, 218
Beta hemolysis, 151, 152
Beverages
  fermented, 132–134
  fermented milk, 131
Bifidobacterium, 199
Binal symmetry, virus, 93
Bioassay, 138–139
Biodegradability, 123
Bioluminescent flagellates, 218
Biosynthesis, 77–84
Bites
  of flies, 155
  of mites, 156

Bites *(cont.)*
  of ticks, 155, 156
Bittner, J., 102
Bivalvulida, 245
"Black death," 155
Black stem rust, 229
Blastocladiella, 223
Blastomyces dermatitidis, 148
Blastomycosis, 148–149, 227
Blastospores, 221
Blattabacterium cuenoti, 202
Blepharisma, 252
Body lice, 156, 181
Boil disease of cyprinid fish, 245
Boils, 150
Bordetella, 146, 188
Borrelia, 156, 181, 182
Botrydium, 213
Botryocella, 240
Bottom yeasts, 133
Botulism poisoning, 145, 195
Bourbon, 134
Bovine anaplasmosis, 203
Bovine spirochetosis, 182
Brachiomonas, 208
Brachyarcus, 182
Branched pathways, control in, 89, 90
Bread, 134
Bread mold, common, 226
Brettanomyces, 132
Brevibacterium, 132
Bright-field microscopy, 15–19
Brown algae, 212
Brucella, 149, 188
Brucellosis, 149
Buboes, 150
Bubonic plague, 155
Budding appendaged bacteria, 179–181
Bunyaviridae, 157
Burkitt's lymphoma, 102
Burnet, MacFarlane, 13
Bursa of Fabricus, 168
Bus (milk beverage), 131
Buttermilk, 131
Butyrivibrio, 118

Cabbage root infections, 223
Calothrix, 124, 178
Calvin-Benson pathway (cycle), 67, 68
Calymmatobacterium, 188
Camembert cheese, 132
Campylobacter, 182
Candida, 130, 132, 138, 228
Candida albicans, 153–154, 228
Candidiasis, vaginal, 153–154
Canine ehrlichiosis, 202

Canker, 198
Cannosphaera, 240
Capsid, virus, 92
Capsomeres, virus, 92
Capsules, bacterial, 39, 141
Carbon cycle, 121–124
Carbon dioxide, *see* CO_2 entries
Carboxysomes, 45
Carchesium, 250
Caries, deep dental, 200
Carnation rust, 229
Carteria, 208
Caryophanon, 195
Caulobacter, 179, 180
Caulerpales, 209
Cedar-apple rust, 229
Cell counts, 33
Cell mass, 35
Cell-mediated immunity, 163, 165, 168
Cell structure
  eucaryotic, 47, 48
  procaryotic, 37–48
Cellular slime molds, 223
Cellulomonas, 198
Cellulose decomposition, 118, 122–123, 194, 198
Cell wall, 40–43
Centrohelida, 241
Ceratium, 218, 219
Cercospora, 228
Chagas' disease, 159, 234
Chain, Ernst, 13
Challengeron, 239, 240
Chamaesiphon, 173, 175
Chamaesiphonales, 175
Chancres, 149
Chara, 210
Characterization media, 29–30
Charophyta, 210
Chase, 14
Chaunacanthida, 238
Cheeses, 131–132, 196, 198, 227
Chemically defined media, 28
Chemolithotrophs, 60, 127
Chemolithotrophy, 67, 69
Chemoorganotrophs, 60
Chemoorganotrophy, 69–77
Chemostat, 54
Chemotherapy, 13
Chemotrophs, 60
Chicken pox, 153
Chilodonella, 249
Chilomonas paramecium, 219
Chlamydiales, 203–205
Chlamydia, 147, 150, 203, 205
Chlamydodon, 249

Chlamydomonas, 111, 112, 207, 208
Chlamydospores, 221
Chloramphenicol, 201
Chlorella, 138, 207, 208
Chlorellales, 208
Chlorobacteraceae, 64
Chlorobiaceae, 64, 174
Chlorobium vesicles, 45
Chlorobotrys, 213
Chloroflexaceae, 64, 174
Chloroglea, 124
Chlorogloeopsis, 178
Chlorohydra viridis, 208
Chloromycetin, 137
Chlorophyll, 64, 175
Chlorophyta, 6, 208–210
Chloroplasts, 47–48, 111, 112
Choanephora cucrubitarum, 226
Choanoflagellida, 215, 232
Cholera, 143
Chonosaurus, 249
Chonotrichida, 249
Chorococcales, 175
Chromatiaceae, 64, 126, 172, 174
Chromatic aberration, 17, 18
Chromatium, 174
Chromobacterium, 124, 138, 188, 189
Chromulinales, 215
Chroococcidiopsis, 177
Chroococcus, 173
Chroomonas, 219
Chrysamoeba, 215
Chrysarchnion, 215
Chrysochromulina, 216
Chrysophyta, 6, 212–218
Chytridiomycota, 6, 221, 223
Chytrids, 223
Ciliophora, 6, 231, 246
Ciliophyrs, 241
Cingulum, 218
Citric acid cycle, 74–75
Citric acid production, 135
Citrus, "stubborn disease" of, 205
Cladosporium, 146
Classification
  Leedale's multikingdom system, 9
  Whittaker's five-kingdom system, 8
Clathrulina, 241
Claviceps purpurea, 226
Closterium, 210, 211
Clostridium, 118, 124, 136, 138, 154, 195
Clostridium botulinum, 130, 142, 145, 192, 195
Clostridium pasteurianum, 80
Clostridium perfringens, 141, 145, 154, 195

Clostridium tetani, 142, 154, 195
Cnidosporidia, 6
CO_2 assimilation, 66–67, 68
CO_2 generation, 75
Coagulase, 151
Coccidia, 242–243
Coccidiosis, 243
Coccodiscus lithelius, 240
Coccolithus pelagicus, 216
Coccus, 37–38
Code redundancy, 104
Codium, 207, 210
Codonella, 252
Codons, 84, 104
Coenocytic hyphae, 220
Coenzyme, 63
Colds, common, 148
Cold sores, 152
Coleps, 246
Colicins, 46
Collagenase, 136, 141, 155
"Collared flagellates," 232
Colony counts, 33
Colpoda, 248, 251
Colpodium, 250
Commensalism, 114
Comomonas, 138
Compatibility, 47
Competent strains, 107
Complementary strands, 98
Complement fixation, 167
Complement system, 161–162
  classical pathway, 162
  properdin pathway, 162
Complex media, 28
Complex symmetry, virus, 93
Concerted feedback inhibition, 90
Concharium, 240
Conidiophore, 221
Conidiospores, 221, 230
Conjugation, 106–107
Conjunctival ulcer, 155
Conjunctivitis, 190
  inclusion, 150, 203
Conoid, 242
Constitutive enzymes, 87
Continuous cultures, 53–54
Coordinate synthesis, 88
Copodida, 248
Core, virus, 92
Corn smut, 230
Corn stunt, 205
Corticosterone, 137
Corynebacterium, 196
Corynebacterium diphtheriae, 100, 142, 146, 196, 197
Cosmarium, 210

Index

Cottage cheese, 132
Coulter counter, 34
Counting chambers, 33
Cowpox, 13
Coxiella burnetii, 147, 202
Coxsackieviruses, 143–144
Craspedomonadales, 215, 234
Cream cheese, 132
Crick, 103
Cristispira, 181
Cryophiles, 57
Cryptococcosis, 149
Cryptococcus, 149, 228
Cryptomonads, 219
Cryptomonas, 219, 233
Cryptophyta, 6, 219
Culture methods, 26–30
    continuous, 53–54
    pure, 12–13
Cultures, starter, 131, 194
Cumulative feedback inhibition, 90
Cup fungi, 227
Curvularia lunata, 137
Cutaneous leishmaniasis, 159
Cyanobacteria, 45, 64, 174–178
Cyanophycin, 44
Cyclic AMP, 143
Cyclic photophosphorylation, 65
Cylindrospermum, 177
Cyprinid fish, boil disease of, 235
Cyprus fever, 149
Cyrtophorida, 249
Cystitis, 196
Cytochrome system, 61
Cytophaga, 138, 178
Cytophagales, 178
Cytotaxins, 161

Dadhi, 131
Dark-field microscopy, 19, 20
Dasycladales, 209
Death phase, 52
Debaryomyces, 130
Decompositions, industrial, 137–138
Defective phage, 108
Delbruck, 103
Deletion mutations, 105
Dengue, 157
Denitrification, 124
Dental caries, deep, 200
Dental plaque, 199, 200
Deoxyribonuclease, 155
Deoxyriboside, 83
Derepression, 88
Dermatophytosis, 153
Dermocarpa, 175, 176

Dermocarpella, 175
Derxia, 185
Desmidium, 210
Desmophyceae, 218
Desmothoraci, 241
Desulfomonas, 126
Desulfotomaculum, 126
Desulfovibrio, 126, 190
Deuteromycetes, 221, 228
Devil's grippe, 144
Dextran, 136
d'Herelle, F., 14
Diagnostic tests, 171
Diatoms, 212, 216
Diauxy, 52–53
Dictyocha, 215, 217
Dictyochales, 215
Dictyosteliia, 237
Dictyostelium, 223
Didinium, 246, 247
Die-away test, 123
Differential staining, 19
Difflugia, 235
Dileptus, 246
Dinobryon, 215, 217
Dinoflagellates, 218
Dinophyceae, 218
Dinophyta, 6
Diphopyridine nuclease, 152
Diphtheria, 146
Diphtheria exotoxins, 165
Diplomonadida, 234
Discomycetes, 227–228
Discorbis, 236, 237
Diseases
    germ theory of, 12
    of humans, 142–160
Disinfection, 32
Distilled liquor, 133–134
DNA, 4, 45–46, 48
    chloroplast, 111, 112
    mitochondrial, 111
    recombinant, 109–110
DNA polymerases, 86, 107
DNA primer molecule, 86
DNA strand separation, 86
DNA structure, 103
DNA synthesis, 86, 98
Domagk, Gerhard, 13
Downy mildews, 224
Dry-heat sterilization, 30
Dry weight, 35
Dunaliella, 208
Dysentery
    amoebic, 144, 235
    bacterial, 142–143, 188
Dysteria, 249

Index

Earth tongues, 227
East African sleeping sickness, 159
Eastern equine enchephalitis, 157
Echoviruses, 144
Ectomycorrhiza, 115
Ectosymbiosis, 115
Eczema herpeticum, 152
Edam cheese, 132
Effectors, 88
Ehrlich, Paul, 13
Ehrlichia canis, 202
Ehrlichiosis, canine, 202
Eimeria, 242, 244
Elaphomycetales, 227
Electron microscopy, 23–26
Electron transport, 61–62, 65
Ellipsoidion, 213
Elphidium, 237
Embden-Meyerhof pathway, 69–71
Emmonsiella capsulata, 227
Empirical media, 28
Emulsion dilution methods, 27
Encephalitis, herpetic, 152
Endemic typhus, 156, 202
Enders, 14
Endogenote, 106
Endomycorrhiza, 115
Endospores, 47
Endosymbiosis, 115, 117, 118, 208, 231
Endosymbiotic algae, 117–118, 208
Endotoxins, 141, 147
End product inhibition, 88
End product repression, 88, 90
Energy generation, 59–64
Energy requirements, 171
Energy storage, 60–62
Enrichment cultures, 29
Entamoeba histolytica, 144, 235
Enteric fevers, 142
Enterobacter aerogenes, 135, 136
Enterobacteriaceae, 188–189
Enterobacterium, 124
Enterotoxins, 142, 143, 145, 194
Entner-Douderoff pathway, 71, 72
Entodiniomorphida, 119, 248
Entodinium, 248
Entry, portals of, 140–141
Enumeration methods, 32–36
Enzyme induction, 88
Enzyme regulation, 14
Enzymes, 62–64
  commercial, 136–137
  invasive, 141, 142
Enzyme synthesis, 87–88
Ephelota, 249, 251
Epidemic typhus, 156
Epidermophyton, 153

Epifluorescence microscope, 33
Epimerite, 242
Episomes, 46
Epistylis, 250
Epstein-Barr virus, 102, 153
Equine encephalitis, 157
Eremothecium ashbyi, 136
Ergot fungus, 226
Ermosphaera, 208
Erwinia, 188–189
Erysipelas, 151
Erysipeloid infections, 196
Erysipelothrix, 195, 196
Erythema nodosum, 152
Erythrogenic toxin, 152
Escherichia coli, 135, 188
  genetic mapping of, 108
Eubacterium, 199
Eucaryotes, 3, 4
Eucaryotic gene expression, 110–113
Eucaryotic microorganisms, 206–252
Eucaryotic organelles, symbiotic origin of, 47–48
Eucoccidiida, 243
Eudorina, 208
Euglena, 211, 212
Euglenophyta, 6, 210–212
Euglypha, 237
Eugregarinida, 242
Eumycetozoea, 220, 223, 230, 237
Eumycota, 220
Euplotes, 251, 252
Eurythermal species, 57
Eustigmatophyta, 6, 213
Exciter filter, 21
Exfoliatin, 151
Exogenote, 106
Exoschizon, 242
Exotoxins, 141
  tetanus and diphtheria, 165
Exponential phase, 51–52
Extreme halophile, 58

F+, 106
F−, 107
fab portion, 163
Fabricus, bursa of, 168
Facultative aerobes, 58
Facultative autotrophs, 60
FAD, 61
fc end, 163
fd virus, 93, 99
Feedback inhibition, 88, 89, 90
Fermentation, 11–12, 69–74
Fermentation end products, 73–74
Fermented beverages, 132–134, 194
Ferredoxin, 65

Index

F factor, 106
Fibrinolysin, 141, 152
Filtration, sterile, 32
Fimbrae, 39
Final dormant phase, 52
Fireblight, 189
Firefly ATP assay, 36
Fischerella, 176, 178
Five-kingdom system of classification, Whittaker's, 8
Flagella, procaryotic, 38–39
Flavobacterium, 188
Flea, rat, 156, 202
Fleming, Alexander, 13
Flies, bites of, 155
Florey, Howard, 13
Fluorescent antibodies, 21, 166
Fluorescent microscopy, 20–21
FMN, 61
Folliculina, 252
Food poisoning, 194
  staphylococcal, 145
  streptococcal, 145
Food spoilage, 129–30, 226, 227
Foraminifera, 237, 238
Forsch, P., 14
Fossil microorganisms, 9
Frame shift mutations, 105
Francisella, 155, 188
Frankia, 200
Frankiaceae, 200
Freeze etching, 24
Frings generator, 134
Frontonia, 250
Fruit spot, 198
Frustrule, 216
Fucus, 212, 214
Fumaric acid, 136
Fungal spores, 221–222
Fungi, 5, 6, 220–230
Fungi Imperfecti, 221, 228
Fusarium, 136, 137, 145, 146, 228

β-Galactosidase, 88, 109
Galactoside permease, 88, 109
Gallionella, 179, 180
Galls, 185
Gangrene, 154, 195
Gargarius, 249
Gaseous autoclave, 30–31
Gastroenteritis, 144
Gas vacuoles, 45
Geleia, 248
Gene expression, eucaryotic, 110–111
Generation time, 49–50
Gene replication, 111

Genetic code, 104
Genetic engineering, 109–110
Genetic mapping, 108
Genophore, 2, 46
Gentamycin complex, 201
Geotrichum, 129, 132
Gephyrocapsa huxleyi, 216
German measles, 148
Germ theory of disease, 12
Giardia, 233, 234
Giardiasis, 144–145
Gibberella fujikuroi, 136
Gibberellic acid, 136
Gigaductus, 242
Gigartacon, 238
Gingivostomatitis, herpetic, 152
Girdle, 216, 218
Glaucoma, 250
Gliding bacteria, 178
Gliding green bacteria, 174
Gloeobacter, 175
Gloeocapsa, 173, 175
Gloeothece, 175
Glomerulonephritis, acute, 151
Gluconic acid production, 135
Glucose-6-phosphate, 71, 72
Glyceraldehyde-3-phosphate, 72
Glycolysis, 69–71
Glyoxylate shunt, 75, 76
Gnotobiology, 155
Golden algae, 212
Golden-brown algae, 213
Golenkinia, 208
Gonium, 207, 208
Gonorrhea, 149, 190
Gonospora, 242
Gonyaulax, 218, 219
Grafts, 13
Grahamella, 203
Gram-negative bacteria, 42
Gram-positive bacteria, 42
Gram stain, 19, 171
Granular reserves, 44–45
Granuloreticulosea, 237
Green algae, life cycles, 209
Green branch, 6
Green gliding bacteria, 64
Green sulfur bacteria, 64, 174
Gregarina, 239, 242
Gregarinia, 242
Grellia, 243
Griseofulvin, 227
Gromia, 237
Growth cycle, 50–53
Growth factors, organic, 82–83
Growth rate, 56
Growth rate constant, 49–50, 55

Gymnodinium, 217, 218, 219
Gymnostomatia, 246

Haemaphysales leporispalustris, 202
Haemophilus, 188
Haemophilus influenzae, 141, 146, 189
"Hairy root" disease, 185
Halimeda, 210
Halobacteriaceae, 186
Halobacterium, 186, 193
Halococcus, 186
Halophile, 58
Halosporidium, 245
Halteria, 252
Hanging drop, 19
Hansenula, 130, 132
Haptens, 163
Haptonema, 215
Haptophyta, 6, 213, 215, 216
Hard cheeses, 131, 132
Haustoria, 116, 117, 220
Hayes, 104
Heavy chains, 163
Helical symmetry, virus, 92
Heliochona, 249
Heliophrya, 249
Heliozoea, 241
Helper phage, 108
Hemagglutinin, 148
Hemiascomycetes, 227
Hemocytometer, 33
Hemolysins, 141, 151, 152
Hemorrhagic septicemia, 189
Henle, Jacob, 12
Hepatitis B, 143
Herpes, 152–153
Herpetic encephalitis, 152
Herpetic gingivostomatitis, 152
Hershey, 14
Hessea, 244
Heterodynamic flagella, 213
Heterokaryosis, 113
Heterokont branch, 6
Heterotrichida, 252
Heterotrophs, 60
Hexacapsula, 246
Hexamita meleagridis, 234
Hexose monophosphate shunt, 71
Hfr, 106, 107
Histone, 110
Histoplasma capsulatum, 149
Histoplasmosis, 149, 227
Holacanthida, 238
Holidic media, 28
Holley, 104
Holophrya, 246

Holosticha, 252
Hooke, Robert, 10
Hops, 133
Hot spots, 105
Humoral immunity, 162, 165
Hyaluronidase, 141, 152, 155
Hydrogen-donating system, 116–117
"Hydrophobia," 157
Hymenomonas carterae, 216
Hymenostomatia, 250
Hypermastigida, 119, 234
Hyphae, 220
Hyphomicrobium, 179, 180
Hyphomonas, 179
Hypocoma, 249
Hypostomatia, 248
Hypotrichida, 252

Ichthyophthirius multifiliis, 250
Icosahedral symmetry, virus, 92
IgA-M, 164–165
Immobilization, 166
Immunity, 13, 162–168
  active, 165
  cell-mediated, 163, 165, 168
  humoral, 162, 165
  passive, 165
Immunoglobulins, 163
  classes of, 164
Impetigo, 151
Inclusion conjunctivitis, 150, 203
Inducible enzymes, 87–88
Industrial decompositions, 137–138
Infection thread, 116
Infectious hepatitis, 143
Influenza, 147–148
Initiation complex, 85
Initiation factors, 85
Inner spore coat, 47
Insecticides, microbial, 137, 195
Insertion mutations, 105
Insertion sequence, 47
Insertion sites, 46
Integrase, 46–47
Interference microscopy, 23
Interferon, 162
Invasive enzymes, 141, 142
Irish late blight of potatoes, 224
Isoenzymes (isozymes), 63, 90
Isolation techniques, 27
Isotricha, 248
Itaconic acid, 136
"Itch," 250
Iwanowsky, D., 13–14

Jablot, Louis, 10–11
Jacob, Francois, 14, 104

Index

Japanese beetle, 202
Jenner, Edward, 13

Kala azar, 160
Karyorelictida, 248
Kefir, 131
Keratoconjunctivitis, 152
Kerona, 252
2-Keto-3-deoxyoctonoic acid, 42
Khorana, 104
Kinetograminophorea, 246
Kinetoplastida, 234
Kingdoms of organisms, 4–5
Kitasatoa purpurea, 201
Klebsiella pneumoniae, 141, 188
Kloeckera, 132
Koch, Robert, 12–13
Koch's postulates, 12, 140
Koumiss, 131
Krebs cycle, 74–75
Kudoa, 246

Labyrinthomorpha (Labyrinthulomycota), 220, 224, 230, 231, 241
Labyrinthula marina, 224
Lacrymaria, 246, 247
Lactic acid, 134, 135, 196
Lactobacillus, 130, 131, 133, 135, 196, 197, 199
Lactobacillus acidophilus, 197
Lactobacillus brevis, 134
Lactobacillus bulgaricus, 131, 135, 196
Lactobacillus lactis, 131, 196
Lactobacillus plantarum, 134, 196
Lactobacteriaceae, 82
Lactose operon, 88
Lag phase, 50–51
Laminaria, 212, 214
Latent phage, 100
Leaching, microbial, 127–128
Leaf spot diseases, 198, 228
Leben, 131
Lecithinase, 141, 154
Lederberg, Joshua, 14, 104
Leedale's multikingdom system of classification, 9
Leghemoglobin, 116
Leishmania, 159, 160
Leishmaniasis, 159–160
Leprosy, 200
Leptomyxida, 237
Leptopharynx, 249
Leptospira, 155, 182
Leptothrix, 179
Leptotrombidium, 202
Leuconostoc, 131, 133, 134, 136, 194

Leukocidin, 141, 151
LHT system, 93–95
Lice, 156, 181, 202
Lichens, 117
Liederkranz cheese, 132
Light chains, 163
Light-harvesting pigments, 64–65
Limestone, 124
Linnaean taxonomy, 7
Lipases, 151
Lipid A, 42
Liptauer cheese, 131
Liquor, distilled, 133–134
Listeria, 195, 196
Litonotus, 246, 247
Loeffler, Friedrich, 14
Logarithmic phase, 51–52
Log death phase, 52
Log phase, 51–52
Lophotrichous flagellation, 39
Loxodes, 247, 248
Loxophyllum, 246
Luminescent organs, 189
Luminous bacteria, 119
"Lumpy jaw," 200
Luria, 103
Lwoff, Andre, 14, 104
Lymphogranuloma venereum, 150, 203
Lyngbya, 177
Lysis, 166
Lysogenic conversion, 100
Lysogenic phage β and λ, 100
Lysogeny, 100, 104
Lysosomes, 161
Lysozyme, 40, 99, 160

McCarty, M., 14, 103, 107
MacLeod, C. M., 14, 103, 107
Macrocyst, 223
Macrocystis, 212
Macroglobulin, 164
Macromolecular synthesis, 84–87
Macrophages, 167
Malaria, 157–158, 242
Malt, 133
Malta fever, 149
Mammary cancer, 102
Manganese bacteria, 191
Mantle, virus, 92
Manual of Clinical Microbiology, 171
Marteilia refringens, 245
Mastigomoneran branch, 5
Mastigophora, 231–235
Mastitis, 196
Mating types, 112
Mattesia, 242
Measles, 148

Index

Mechanical barriers, 160
Medawar, Peter, 13
Meiosis, 111, 112
Membrane filter counts, 34
Membrane filtration method, 35
Meningitis, 146
  aseptic, 143, 144
  meningococcal, 147
  septic, 147
Meningococcal meningitis, 147
Meningoencephalitis, 144, 152
Merozygotes, 106
Mesocaryotic nuclei, 218
Meso-diaminopimelic acid, 40
Mesophiles, 57
Mesosomes, 44, 46
Messenger RNA, 84
Metabolic pathways, 78
Metabolic regulation, 87–90
Metachromatic granules, 44
Metchnikoff, Elie, 13
Metchnikovella, 244
Methane-oxidizing bacteria, 186
Methane production, 138
Methanobacteriaceae, 193
Methanobacterium, 123, 192
Methanococcus, 123
Methanogenesis, 123
Methanogenic bacteria, 193
Methanogens, 193
Methanomonas, 186
Methanosarcina, 123
Methylobacter, 186
Methylobacterium, 186
Methylococcus, 186
Methylocystis, 186, 187
Methylomonadaceae, 186
Methylomonas, 183, 186
Methylosinus, 186
Metopus, 252
Micrasterias, 210, 211
Microaerophiles, 58
Microbial activities in biosphere, 120–128
Microbial insecticides, 137, 195
Microbial leaching, 127–128
Microbial steroid transformation, 137
Microbiology, vii
  applied and industrial, 129–139
  history and development of, 10–14
Micrococcaceae, 193–194
Micrococcus, 130, 135, 138, 193–194
Micromonosporaceae, 201
Micronutrients, 82
Microorganisms, 1–2
  attenuated living, 165
  classification of, 4–9

Microorganisms *(cont.)*
  early evolution of, 9–10
  eucaryotic, 206–252
  genetics of, 103–113
  major categories of, 2–4
  phagocytosis of, 161
Microscopy, 15–26
  bright-field, 15–19
  dark-field, 19, 20
  electron, *see* Electron microscopy
  epifluorescence, 33
  fluorescent, 20–21
  interference, 23
  phase-contrast, 21–22
  ultraviolet, 19–21
Microspora, 244
Microsporea, 231, 244
Microsporidia, 244
Microsporum, 153
Microthorax, 249
Milk beverages, fermented, 131, 194
Milk products, 130–132
Minchinia, 245
Mineralization, 124
Minisporida, 244
Mites, bites of, 156
Mitochondria, 47–48, 111
Mitochondrial genetics, 111–112
Mitochondrial mutants, 112
Mitosis, 111
Modification enzymes, 99
Monera, 5
Monocystis, 239, 242
Monod, Jacques, 14, 88, 104
Monoxenic culture, 28
Moraxella, 124, 190
Morels, 226, 227–228
Mosquito, *Anopheles*, 157–158
Most probable number (MPN), 34
Mucocutaneous leishmaniasis, 159–160
Mucron, 242
Multivalvulida, 246
Mumps, 148
Murein, 2
Murine typhus, 202
Mushroom caps, 222
Mushrooms, 228
Mutagenic agents, 105–106
Mutagens, 106
Mutations, 105–106
Mutualism, 114
Mycelia, 220
Mycobacteriaceae, 200
Mycobacterium tuberculosis, 146, 200
Mycoplasma, 204, 205
Mycorrhiza, 115–116
Mycotoxicosis, 145

Mycotoxin, 145
Mylestoma, 252
Myocarditis, 144
Myxobacterales, 178
Myxobolus, 245, 247
Myxogastria, 224, 237
Myxomoneran branch, 5
Myxomycetes, 237
Myxomycota, 6, 221, 224
Myxosarcina, 177
Myxosporea, 245
Myxosporean, life cycle of, 245
Myxospores, 178
Myxozoa, 231, 245

NAD, 61
NADH$_2$, 75
NADP, 61
Naegleria, 144, 235
Nannizzia, 227
Nasopharyngitis, 151
Nassellarida, 240
Nassula, 249
Nassulida, 248–249
Natural resistance, 160–162
Necrotic wilt disease, 189
Needham, John, 11
Negative growth acceleration phase, 52
Negative staining, 19
Neisseria gonorrhoeae, 149, 187, 190
Neisseria meningitidis, 147, 190
Nematode trapping fungi, 226
Neogregarinida, 242
Neonatal herpes, 153
Neonocardin, 201
Nephelometry, 35
Nereocystis, 212
Neufchâtel cheese, 132
Neuraminidase, 147–148
Neurospora, 129, 225
Neurospora crassa, 103, 112–13, 227
Neurotoxin, 142, 154
Neutralism, 114
Neutralization, 166
Nevskia, 179
Nirenberg, 104
Nitella, 210, 211
Nitrate assimilation, 79
Nitrate reduction, assimilatory, 125
Nitrification, 125–126
Nitrifying bacteria, 126, 191
Nitrite assimilation, 79
Nitrobacter, 126, 191
Nitrobacteraceae, 126, 191
Nitrobacter winogradskyi, 192
Nitrococcus, 126, 191
Nitrogen, total, 36

Nitrogen-activating system, 117
Nitrogenase, 116
Nitrogen assimilation, 80–81
Nitrogen cycle, 124–126, 190, 191
Nitrogen fixation, 116–117, 124, 185, 188, 200
Nitrosococcus, 187, 191
Nitrosofying bacteria, 126, 191
Nitrosolobus, 126, 191
Nitrosomonas, 126, 191
Nitrosospira, 126, 191
Nitrospina, 126, 191
Nocardia, 201
Nocardiaceae, 201
Nocardia kuroishii, 201
Noctiluca, 218
Nodularia, 177
Noncyclic photophosphorylation, 65–66
Nonsense codons, 85, 104, 105
Nosema, 244, 247
Nosema locustae, 137
Nostoc, 117, 124, 176, 177
Nostocales, 177
Nucleocapsid, virus, 92
Nucleoid
  procaryotic, 45–46
  virus, 92
Nucleoside, 83
Nucleosomes, 110
Nucleotide, 84
Numerical aperture, 16
Numerical taxonomy, 9
Nutrient concentration, 55, 56
Nyctotherus, 252

Obligate aerobes, 58
Obligate halophiles, 58
Occulosporida, 245
Oceanospirillum, 182
Ochratoxins, 145
Ochromonadales, 215
Ochromonas, 214, 215
Octodendron, 240
Odontostomatida, 252
Oidium, 221
Oligidic media, 28
Oligohymenophorea, 249–250
Oligotrichida, 252
Olpidium brassicae, 223
Oncogenic viruses, 100–102
One gene–one enzyme theory, 14, 103
Oomycota, 221, 223–224, 241
Oosphere, 222
Oospores, 222
Opalina, 236
Opalinata, 231, 235
Operator gene, 109

Operon concept, 104, 108–109
Ophryoscolex, 248
Opsonins, 161
Opsonization, 166
Oral thrush, 153
Organic solvent production, 136
Orleans process, 134
Ornithodoros, 181
Orotic acid, 84
Oroya fever, 203
Orthodonella, 248
Orthomyxoviruses, 147
Oscillatoria, 176, 177
O side chain, 42
Osmophilic microorganisms, 57
Osmotolerant microorganisms, 57
Osteomyelitis, 151
Ouchterlony technique, 166
Outer spore coat, 47
Overlapping genes, 104
Oxymonadida, 235
Oxytricha, 252

Pandemics, 148
Pandorina, 208
Papilloma, 102
Paracoccus, 190
Parainfluenza viruses, 148
Paralytic poliomyelitis, 143
Paramecium, 117, 208, 250, 251
Paramylon, 210
Paramyxa, 245
Paramyxea, 245
Paramyxoviridae, 148
Parasitism, 114
Paratrophs, 60
Pascherina, 208
Pasteur, Louis, 11, 13
Pasteur effect, 77
Pasteurella, 155, 188, 189
Pasteurization, 12, 130
Pathogenicity, mechanisms of, 141–142
Peach leaf curl disease, 227
Pebrine disease, 244
Pediculus humanus, 156, 181, 202
Pediococcus, 133, 194
Pedomicrobium, 179
Pelobiontida, 235
Pelomyxa, 235
Pelosigma, 182
Penicillin, 13, 137, 227
Penicillinase, 151
Penicillium, 129, 132, 135, 137, 145, 225, 227
Penicillium camemberti, 132
Penicillus, 210
Pentatrichomonas hominis, 234

Peptidoglycan, 2, 40–41
Peptidyl transfer, 85
Peptococcaceae, 194
Petococcus, 194
Peptostreptococcus, 194
Peranema, 212, 214
Pericarditis, 144
Peridinium, 217
Periodontal disease, 200
Peritrichida, 250
Peritrichous flagellation, 39
Perkinsea, 241
Perkinsus, 242
Permeases, 43
Petrefaction, 125–126
Petri, R. J., 13
Phaeocalpida, 240
Phaeoconchida, 240
Phaeocystida, 240
Phaeodaria, 240
Phaeodendrida, 240–241
Phaeogromida, 240
Phaeophyta, 212
Phaeosphaerida, 240
Phage conversion, 100
Phage integration, 100
Phagocytes, types of, 161
Phagocytosis of microorganisms, 161
Phagolysosome, 161
Phase-contrast microscopy, 21–22
Phenol coefficient, 32
θX 174 virus, 104
Phlebitinys verrucarum, 203
Phlebotomus, 159
Phormidium, 177
Phosphoketolase pathway, 73
Phosphorus assimilation, 82
Phosphorylation, substrate level, 61
Photobacterium, 119, 189
Photoblepharon, 119
Photolithotrophs, 60, 67
Photoorganotrophs, 60, 67
Photophosphorylation
  cyclic, 65
  noncyclic, 65–66
Photosynthetic pathway, 68
Photosynthetic pigments, 65
Photosynthetic reaction center, 65
Photosystem I pigments, 65–66
Photosystem II pigments, 65–66
Phototrophic bacteria, 172–174
Phototrophs, 60
Phototrophy, 64–67
Phycobiliproteins, 175
Phycocyanin, 175
Phycoerythrins, 175
Physarum, 225

Physarum polysephalum, 224
Physoderma, 223
Phytomastogophorea, 231–232
Phytophthora infestans, 224
Pichia, 130
Pickles, 134, 196
Pigment absorption, 35
Pili, 39
Piroplasmia, 242, 244
Plague, 189
Planomonospora, 197, 201
Plantae, 5
Plant pathogens, 189, 198, 205, 226, 228, 229
Plant tumors, 185
Plaque assay, 96
Plaque-forming unit, 96
Plasma cells, 167
Plasma membrane, procaryotic, 43–44
Plasmid genes, 142
Plasmids, 46–47, 100, 106, 109
Plasmodiophorea, 220, 230, 237
Plasmodiophoromycota, 6
Plasmodium, 157, 242, 244, 247
Plasmopara viticola, 224
Plectomycetes, 227
Plectonema, 177
Pleodorina, 208
Pleurochloris, 213
Pleuropneumonia, 205
Pleurostomatida, 246
Pneumococcal pneumonia, 146–147
Pneumonia, 146–147, 188
Podophyra, 249
Point mutations, 105
Poisoning, food, 145, 195
Polar flagellation, 39
Polio vaccines, 14
Poliovirus, 143
Poly-β-hydroxybutyric acid, 44
Polycystinea, 240
Polyhedral bodies, 45
Polyhymenophorea, 252
Polypeptide folding, 86
Polyplax spinulosus, 202
Polyribosomes, 44
Polyspondylium, 223
Polyvalent antigens, 166
Popillia, 202
Porphyridium, 217
Portals of entry, 140–141
Positive acceleration phase, 51
Potato spindle tuber disease, 93
Potato wart, 223
Pour-plate methods, 27
Powdery mildews, 227
Precipitation, 166

Preinfection stage, 116
Primary production, 122
Primary response, 167
Primary stimulus, 167
Primociliatida, 246
Procaryotes, 2–4
  cell structure, 37–48
  genetics, 103–110
Promoter gene, 109
Promoters, 87
Promycelium, 230
Properdin pathway, 162
Prophages, 100, 142
Propionibacteriaceae, 198–199
Propionibacterium, 132, 198, 199
Propionibacterium freudenreichii, 136, 198
Prorocentrum, 218, 219
Prorodon, 246
Prosthecae, 39–40, 179
Prosthecomicrobium, 179
Prosthetic group, 63
Prostomatida, 246
Protein, single cell, 138
Proteinase, 155
Protein synthesis, 84–86, 104
Protermonadida, 235
Proteus, 188
Protista, 5–6
Protococcidiida, 243
Protoplasts, 41
Protosteliia, 237
Protosteliomycetes, 220, 237
Protozoa, 6, 230–252
  correspondence to algal groups, 232
Prymnesiophyceae, 215, 216
Prymnesium, 216, 217
Pseudoanabaena, 177
Pseudomonadaceae, 184
Pseudomonas, 124, 130, 131, 138, 180, 183, 184
Pseudomonas aeruginosa, 184
Pseudoplasmodium, 223
Psittacosis-ornithosis, 147
Psychrophiles, 57
Puccina graminis, 229
Puerperal fever, 151
Puffballs, 222, 228
Pulmonary histoplasmosis, 149
Pulse labeling, 78
Punctuation in genetic coding, 104
Pure culture methods, 12–13
Purine synthesis, 83–84
Purple nonsulfur bacteria, 64, 172
Purple sulfur bacteria, 64, 172, 174
Pyelonephritis, 196
Pyrenomycetes, 227

Pyrimidine synthesis, 83–84
Pyrocystis, 218
Pyrodinium, 218
Pyrrhophyta, 6, 218–219
Pyruvic acid, 72

Q fever, 147, 202
Quinqueloculina, 237

r II gene, 105
Raabenia, 248
Rabbit ticks, 202
Rabdiophyrys, 241
Rabies, 157
Radiation, sterilization by, 32
Raphidophyta, 213
Rat-bite fever, 189
R core, 42
r-determinants, 46
Recognition, 85
Recombinant DNA, 109–110
Recombination, bacterial, 14, 106–108
Red algae, 219
Red bakery mold, 227
Redi, Francesco, 10
Red leg disease, 189
Red tides, 219
Red-water fever, 244
Refractive index, 16
Relapsing fever, 156–157, 181
Rennet, 131
Replicative form, 98
Replicator site, 46
Replicon, 106
Repression, end product, 88, 89
Repressor gene, 109
Repressors, 88
Resistance
  acquired, 162–168
  natural, 160–162
Resistance transfer factor (RTF), 46
Resolution, 15–17
Respiration, 75–77
Respiratory deficient mutant, 111
Respiratory infections, 148
Restriction endonucleases, 109
Restriction enzymes, 99
Reticuloendothelial system, 161
Retortomonadida, 235
Retroinhibition, 88
Retting, 137
Reverse transcription, 98
Rheumatic fever, 151–152
Rhinoviruses, 148
Rhipicephalus sanguineus, 202
Rhizobiaceae, 185
Rhizobium, 116, 183, 185

Rhizoids, 220
Rhizopodea, 235
Rhizopus, 129, 135, 137, 225, 226
Rhizopus nigricans, 136
Rhodomicrobium, 172, 173
Rhodophyta, 219–220
Rhodospirillaceae, 64, 172, 174
Rhodospirillum, 172, 173
Rhodotorula, 130, 228
Rhynchodida, 249
Rhytidocystis, 243
Ribose-5-phosphate, 71
Riboside, 83
Ribosomes, 4, 44, 85
Ribulose diphosphate carboxylase, 67
Rice water stools, 143
Rickettsiaceae, 202
Rickettsia, 156, 202
Rickettsiales, 201–203
Rickettsial pox, 156
Rickettsia rickettsii, 202
Rickettsia tsutsugamushi, 156, 202
Rickettsia typhi, 202
Ringworm, 153, 227
Ripening cheeses, 132, 198, 227
RNA
  messenger, 84, 104, 111
  transfer, 84–85, 111
RNA polymerization, 87
RNA replicase, 98
RNA synthesis, 87, 98
RNA triplets, 104
Robbins, 14
Rocky Mountain spotted fever, 156, 202
Rodent mites, 156
Rolling circle replication, 100
Root rot of grapevines, 227
Roquefort cheese, 131
Rosellina nectrix, 227
Rothia dentocariosa, 197, 200
Rotulopsis, 132
Rous, Peyton, 14
RTF (resistance transfer factor), 46
Rubella, 148
Rubeola, 148
Rudimicrosporea, 244
Rumen bacteria, 190, 194, 195
Rumen ciliates, 248
Ruminant symbiosis, 118–119
Ruminococcus, 118, 194
Rusts, 221, 229

Saccharomyces, 130, 132
Saccharomyces carlsbergensis, 133
Saccharomyces cerevisiae, 111–112, 133, 134, 225, 227
Sake, 133

Index

Salmonella, 42, 142, 144, 188
Salmonella typhi, 142, 187
Salmonella typhimurium, 142
Saprodinium, 252
Saprolegnia, 224
Sarcodina, 6, 231, 235
Sarcomastigophora, 231-241
Sargassum, 212
Sauerkraut, 196
Saxitoxin, 219
Scanning electron microscope, 24-26
Scenedesmus, 138, 208
Schizoema, 218
Schizopyrenida, 235
Schulze, Franz, 11
Schwann, Theodor, 11
Sclerotium, 224
Scotch whiskey, 133
Scrub typhus, 156, 202
Scuticociliatida, 250
Scytonema, 177
Sea lettuce, 208
Seaweeds, 209, 210, 212, 219
Secondary reaction, 167
Sedgewick-Rafter counting chamber, 33
Selective inhibition, 29
Selective media, 28-29
Selenidium, 242
Selenomonas, 190
Septate hyphae, 220
Septic meningitis, 147
Sequential feedback inhibition, 90
Serology, 165
Serratia, 130, 188
Serum hepatitis, 143
Sewage treatment, 137-138
Sex pilus, 107
Sheathed bacteria, 179
Shigella, 188
Shigella dysenteriae, 142
Shingles, 153
Shope, R., 102
Silent mutations, 105
Single cell protein, 138
Siphonocladales, 209
Sirenin, 223
Skin-sensitizing antibody, 164
Sleeping sickness, 234
Slime layers, 39
Slime molds, 223, 224
Smallpox, 13, 153
Smuts, 229, 230, 231
Snapdragon rust, 229
Soil bacteria, 201
Sonderia, 248
Soredia, 117

Sorocarp, 223
Spallanzani, Lazzaro, 11
Spanish olives, 134
Species concept, 7-9
Sphaerotilus, 176, 179
Sphenophyra, 249
Spherical aberration, 17
Spheroplasts, 41
Spiral bacteria, 182-184
Spirillum, 38, 124, 182, 183
Spirochaeta, 180, 181
Spirochetes, 181-182
Spirogyra, 210, 211
Spiroplasma citri, 205
Spirostomum, 252
Spirotrichonympha, 234
Spirulina, 138, 177
Spontaneous generation controversy, 10-11
Spontaneous mutation, 103
Sporangiomycin, 201
Sporangiophore, 221
Sporangiospores, 221
Sporangium, 221
Spore core, 47
Spore cortex, 47
Spores, fungal, 221-222
Sporothrix schenckii, 155
Sporotrichosis, 155
Sporozoa, 6
Sporozoea, 242
Spumellarida, 240
Staining, 17, 19
Standard plate count, 34-35
Stanley, Wendell, 14
Staphylococcal food poisoning, 145
Staphylococcal hemolysins, 151
Staphylococcus, 130, 194
Staphylococcus aureus, 141, 145, 150-151, 192, 194
Staphylokinase, 151
Starter cultures, 131, 194
Stationary phase, 52
Staurastrum, 210
Stellatosporea, 244
Stenothermal species, 57
Stentor, 252
Stephanopogon, 246
Stereomyxida, 237
Sterigma, 228
Sterile filtration, 32
Sterile media, 27
Sterilization methods, 30-32
Steroid transformation, microbial, 137
Sticholonche, 241
Stigonematales, 178
Streptobacillus, 188, 189

Streptococcaceae, 194
Streptococcal food poisoning, 145
Streptococci, 151–152
Streptococcus, 131, 132, 136, 151, 194
Streptococcus faecalis, 145, 151
Streptococcus lactis, 130, 131, 194
Streptococcus pneumoniae, 141, 146, 194
Streptococcus pyogenes, 151, 192, 194
Streptodornase, 136, 152
Streptokinase, 136, 152
Streptomyces, 113, 136, 137, 197, 201
Streptomycetaceae, 201
Streptomycin, 137
Strobilidium, 252
Strombidium, 252
Structural genes, 88
"Stubborn disease" of citrus, 205
Stylochona, 249
Stylonychia, 252
Substrate, 62
Substrate level phosphorylation, 61
Suctoria, 249
Sulcus, 218
Sulfanilamide, 13
Sulfolobus, 186, 191, 193
Sulfur assimilation, 81
Sulfur cycle, 126, 127, 191
Summer grippe, 144
Supplementary reading, 253–263
Surface spreading methods, 27
Swiss cheese, 132, 198
Symbiodinium microadriaticum, 219
Symbiosis, 114–119
Symbiotes lectularius, 202
Symbiotic origin of eucaryotic organelles, 47–48
Symmetric mode replication, 98
Symphyacanthida, 238
Synchronous growth, 53
Synchytrium endibioticum, 223
Synechococcus, 173, 175
Synechocystis, 175
Synhymeniida, 248
Synthetic media, 28
Synura, 214, 215
Synxenic culture, 28
Syphilis, 149–150

Takadiastase, 136–137
Taphrina deformans, 227
Tartar, dental, 199
Tatum, Edward, 14, 103, 104
Taxonomy, 4, 7, 9
Taxopodia, 241
T cells, 163, 167–168
Teichoic acids, 41–42

Telia, 229
Teliomycetes, 229
Teliospores, 222, 229, 230
Temin, H. M., 102
Temperate phage, 100
Temperature, 55, 57
Termination-release, 85–86
Termite flagellates, 234
Tetanospasmin, 154
Tetanus, 154, 165, 195
Tetrahymena, 250
Tetramitus, 235, 236
Thalassiocolla, 239, 240
Thallus, 220
Theleria, 244
Thermophiles, 57
Thermoplasma, 193
Thermus, 188
Thiobacillus, 126, 127, 191
Thiobacterium, 126
Thiodendron, 179
Thiorhodaceae, 64
Thiospira, 126
Thiothrix, 178
Thrush, oral, 153
Thylakoids, cyanobacterial, 45
Ticks, bites of, 155, 156
Tillina, 248
Tinea, 153
Tintinnopsis, 252
TMV (tobacco mosaic virus), 14, 92
Toadstools, 228
Togaviridae, 157
Togavirus, 148
Tokophrya, 249
Tonsillitis, 151
Top yeasts, 133
Torulopsis, 130, 228
Toxins, 141–142
Toxoids, 165
Toxoplasma, 242, 244
Toxoplasmosis, 244
Trace elements, 82
Tracer methods, 78
Trachelocerca, 248
Trachelomas, 212
Trachoma, 150, 203
Transcription, 84, 87
 reverse, 98
Transduction, 14, 108
Transformation, 14, 101, 103, 107
Transition, 105
Translation, 84–85
Translocation, 85
Transmission electron microscope, 24, 25
Transposons, 47

Index

Transversion, 105
Trebouxia, 117
Trench fever, 156, 202
Trentepohlia, 117
Treponema, 150, 181
Treponema pallidum, 149–150, 181
Triactinomyxon ignotum, 246
Tribonema, 213
Tricarboxylic acid cycle, 74–75
Trichodina, 247, 250
Trichodiniasis, 250
Trichomonadida, 119, 234
Trichomonas, 154, 233, 234
Trichomoniasis, 154
Trichomycetes, 226
Trichonympha, 234, 236
Trichophyton, 153
Trichosida, 235
Trichosphaerium, 235
Trichosporon cutaneum, 228
Trichostomatida, 248
Tridacnia, 219
Tritrichomonas foetus, 234
Truffles, 226, 227
Trypanosoma, 159, 233, 234
Tsetse flies, 159
Tuberculosis, 146, 200
Tularemia, 155–156
Tumor viruses, 100–102
Turbidometry, 35, 36
Turbidostat, 54
Tuscarora, 240
Twort, F. W., 14
Tyndall, John, 11
Typhoidal tularemia, 156
Typhoid fever, 142, 188
Typhus
   endemic, 156, 202
   epidemic, 156
   murine, 202
   scrub, 156, 202

Ulceroglandular tularemia, 155
Ulcero-ocular tularemia, 155
Ulothrix, 209
Ultraviolet microscopy, 19–21
Ulva lactuca, 207
Ulvales, 208–209
Undulant fever, 149
Unicapsula, 246, 247
Urceolaria, 250
Uredinales, 229
Uredinia, 229
Urediniospores, 222, 229
Urosporidium, 245
Urostyla, 252

Ustilaginales, 230
Ustilago maydis, 230

Vaccines, 165
Vaginal candidiasis, 153–154
"Vaginal itch," 234
Vahlkampfia, 235
Vampyrella, 237
Van Leeuwenhoek, Anton, 10
Varicella, 153
Variola, 153
Vaucheria, 213
Veillonella parvula, 190
Verruca peruana, 203
Vestibuliferia, 248
Vibrio cholerae, 142, 143, 187, 189
Vibrionaceae, 188, 189
Vibrio parahaemolyticus, 189
Vinegar, 134
Virion, 92
Viroids, 93
Virulence, 141
Viruses, 2, 3, 91–102
   adsorption, 97
   assembly, 99
   attachment, 97
   binal symmetry, 93
   capsid, 92
   capsomeres, 92
   classification, 93–95
   core, 92
   cultivation, 97
   discovery, 13–14
   enumeration, 96
   general properties, 91
   isolation, 96–97
   late synthesis, 99
   liberation, 99
   mantle, 92
   maturation, 99
   nucleic acid replication, 98
   nucleocapsid, 92
   nucleoid, 92
   oncogenic, 100–102
   origin, 102
   penetration, 97–98
   replication, 97–99
   size, 96
   symmetry, 92–93
   transcription, 98
   tumor, 100–102
Visceral leishmaniasis, 160
Vischeria, 213
Vitamin production, 136
Volutin, 44
Volvocales, 208

Volvox, 207, 208
Vorticella, 250, 251

Waksman, Selman, 13
Wart (filtrate), 133
Warts, 153
"Wasting disease" of eel grass, 224
Watson, 103
Weller, 14
West African sleeping sickness, 159
West Nile fever, 157
Wet mount, 19
Wheat rust, 229
Whiskey, 133
White piedra, 228
"White spot," 250
Whittaker's system of classification, 8
Whooping cough, 146
Wilkins, 103
Wilt, 228
Wine, 132–133
Woodruffia, 248
"Woolly knot" disease, 185
Wormy halibut, 246

Xanthomonas, 124

Xanthophyta, 6, 212, 213
Xenococcus, 175
Xenophyophorea, 237
Xenopsylla cheopis, 202
Xylulose-5-phosphate, 73

Yaws, 150
Yeasts, 113, 228
 basidiomycetous, 229
Yellow fever, 157
Yellow-green algae, 213
Yersinia pestis, 155, 187, 189
Yogurt, 131, 196

Zinder, 104
Zoochlorellae, 117
Zoomastigina, 6
Zoomastigophorea, 232
Zoopagales, 226
Zoospores, 206, 208, 221
Zoothamnium, 250
Zoster, 153
Zygnematales, 210
Zygomycota, 6, 221, 224–226
Zygospores, 222, 224
Zymomonas, 188